半参数模型的理论与应用

魏传华　著

科学出版社

北　京

内 容 简 介

近三十年来，半参数模型在统计学和计量经济学领域里得到了广泛关注和深入研究. 本书主要研究了部分线性变系数模型和半参数可加模型在测量误差或缺失数据等复杂情形下的估计和检验问题，此外，还研究了部分线性模型的检验问题和非参数协方差分析模型.

本书可供统计学和经济学等相关专业的科研人员和学生使用.

图书在版编目(CIP)数据

半参数模型的理论与应用/魏传华著. —北京：科学出版社，2013.11

ISBN 978-7-03-038924-4

Ⅰ. ①半… Ⅱ. ①魏… Ⅲ. ①半参数模型-研究 Ⅳ. ①O211.3

中国版本图书馆 CIP 数据核字（2013）第 248138 号

责任编辑：陈玉琢／责任校对：宣 慧

责任印制：赵德静／封面设计：王 浩

科学出版社 出版

北京东黄城根北街 16 号

邮政编码：100717

http://www.sciencep.com

北京凌奇印刷有限责任公司 印刷

科学出版社发行 各地新华书店经销

*

2013 年 11 月第 一 版 开本：720×1000 1/16

2013 年 11 月第一次印刷 印张：10

字数:200 000

POD定价： 56.00元

（如有印装质量问题，我社负责调换）

前　言

实际问题研究中, 我们往往需要探求变量之间的关系. 回归分析作为解决此类问题的一个有效统计分析方法而得到广泛的应用. 毫不夸张地说, 相关与回归分析(特别是线性回归分析) 是除了描述性统计方法之外应用最为广泛的统计分析方法. 同时, 回归分析也是数理统计领域一个内容极其丰富，得到研究较多的分支.

近三十年来, 随着计算机技术的飞速发展, 越来越多的半参数模型被提出用以探求因变量和自变量之间蕴涵的复杂关系. 本书主要针对部分线性变系数模型和部分线性可加模型，以及部分线性模型和非参数协方差分析模型这几类半参数模型的估计和检验问题做深入的探讨.

第 2 章 ~ 第 6 章研究了部分线性变系数模型的推断问题. 第 2 章首先提出了一种 profile Lagrange 乘子方法用以检验参数分量的线性约束条件, 其次构造了参数分量的 backfitting 估计. 第 3 章研究了异方差情形下的部分线性变系数模型, 讨论了参数分量的有效估计量的构造问题, 并提出了检验异方差的 profile 得分方法. 第 4 章针对一类部分线性变系数变量含误差模型构造了参数分量的约束估计, 并提出了针对约束条件的检验方法. 第 5 章讨论了当非参数部分的协变量不能精确观测时部分线性变系数模型的估计问题. 第 6 章讨论了部分线性变系数模型在因变量存在缺失和线性部分自变量不能精确观测时的估计问题. 第 7 章和第 8 章主要研究了部分线性可加模型. 第 7 章构造了模型的 profile 最小二乘估计及相应的约束估计, 同时提出了针对参数分量的广义似然比检验方法. 第 8 章则讨论了一类部分线性可加变量含误差模型的经验似然推断. 第 9 章主要讨论了几种利用部分线性模型检验线性关系的方法. 第 10 章则对非参数回归曲线的比较问题进行了深入研究.

本书内容大都已在国内外学术期刊上公开发表, 衷心感谢投稿过程中评审人的宝贵意见. 此外, 本书的第 2 章、第 3 章、第 9 章和第 10 章是作者博士学位论文中的内容. 衷心感谢导师吴喜之教授在我读博期间给予的指导和帮助以及我工作后一直给予的鼓励和支持. 感谢科学出版社陈玉琢编辑在本书出版过程中给予的帮助.

由于作者水平有限, 书中难免有不妥之处, 恳请同行和广大读者批评指正.

魏传华

2013 年 6 月

中央民族大学

目　　录

第1章　介　　绍

1.1　引　　言

实际数据分析中，大量问题涉及两组变量间的关系的研究. 回归分析作为解决此类问题的一个有效的统计方法而得到十分广泛的应用. “回归 (regression)”一词是由英国著名生物学家和统计学家 F.Galton 在研究人类遗传问题时提出的. 从早期的一元及多元线性回归模型，到非线性回归模型，以及后来为处理离散变量发展起来的广义线性模型，回归分析一直是数理统计学研究的最热点之一. 多样化的回归模型及其拟合方法构成了数理统计学的一个内容极其丰富，应用十分广泛的分支.

设因变量为 Y，自变量为 $X_1, X_2, \cdots, X_p$，回归分析的主要内容就是探求因变量与自变量之间的关系，即建立如下的模型

$$Y = f(X_1, X_2, \cdots, X_p) + \varepsilon, \tag{1.1}$$

其中 ε 为误差项，$f(\cdot)$ 称为回归函数. 回归分析的主要任务包括基于 Y 和 $X_1, X_2, \cdots, X_p$ 的观测值对回归函数 $f(X_1, X_2, \cdots, X_p)$ 做估计，选择出对因变量有显著影响的自变量等，以建立 Y 与 $X_1, X_2, \cdots, X_p$ 之间一个有效的关系，从而基于该关系可以度量变量之间的相互影响程度以及用于预测.

回归模型的种类非常多. 根据回归函数 $f(\cdot)$ 的形式，可以简单地分为线性回归模型和非线性回归模型. 线性回归模型一般可记为如下的形式

$$Y = \beta_0 + X_1\beta_1 + X_2\beta_2 + \cdots + X_p\beta_p + \varepsilon, \tag{1.2}$$

其中 $\beta_1, \beta_2, \cdots, \beta_p$ 为未知参数. 该模型已经得到广泛的研究，从参数的估计、自变量的选取、模型的诊断分析与统计推断等各个方面，已建立起了一套完整的理论与方法体系并得到广泛的应用，有关的内容可参考 Rao (1973), Ryan (1997) 和 Rao 等 (2008) 等. 非线性回归模型可记为如下形式

$$Y = g(X_1, X_2, \cdots, X_p; \beta_1, \cdots, \beta_p), \tag{1.3}$$

其中 $g(.)$ 为某个关于未知参数 $\theta_1, \cdots, \theta_p$ 非线性的已知函数. 此类回归模型无论从应用还是从理论都得到了较为充分的研究，形成了较完整的体系，可见 Bates 和 Watts (1988), Seber 和 Wild (1989) 等.

当因变量 Y 为 0-1 分布、Possion 分布等离散分布时, 直接用最小二乘法拟合形式 (1.2) 的回归模型是不可行的. 为了解决这一问题, 20 世纪 70 年代由 J.A.Nelder 等提出并深入研究了广义线性模型, 这类模型将因变量的分布推广到了指数族分布. Logistic 回归模型和 Possion 回归模型就是得到广泛应用的两类广义线性模型.

线性回归模型和非线性回归模型以及广义线性模型都属于参数模型, 即它们共同的特点是因变量和自变量之间的函数关系是指定的特定形式. 正如梅长林 (2000) 所说, 参数回归模型有很多优点. 首先, 模型中的参数一般具有某种物理意义, 从而有利于对分析结果的解释. 事实上, 在许多回归分析的问题中, 人们对参数的兴趣胜过回归函数本身. 在经济学中应用最为广泛的 C-D 生产函数模型可以转化为如下形式的线性回归模型

$$\ln Y = \ln A + \alpha \ln K + \beta \ln L + \varepsilon,$$

其中 Y, K, L 分别是产出、资本投入和劳动投入. 模型系数 α 和 β 在经济学中的含义分别是资本和劳动的产出弹性. 其次, 参数回归模型易于统计分析, 以线性回归模型为例, 显然对于回归函数的研究就转化为对几个参数 $\beta_1, \cdots, \beta_p$ 的研究上. 最后, 当我们所设定的参数模型形式与实际数据比较吻合时, 显然对整个回归函数的估计在统计上具有更高的有效性, 分析结果更具有针对性和说服力.

实际数据分析中, 因变量与自变量之间到底是什么样的关系? 我们需要从实际数据以及问题专业背景出发进行探索, 而不是为了推断上的简便而将这种关系简单地假定为线性或者其他具体的形式. 在大多数实际问题中, 人们对回归函数的有关信息知之甚少, 通常对其具体形式的假定具有很大的主观性. 设定不当的参数回归模型有可能导致对回归曲线 (或曲面) 的不符合实际的甚至是完全错误的推断. 此外, 对于越来越多的复杂数据, 用某一特定形式的参数回归模型去分析显然是不能够刻画其复杂的变化规律. 从上面的分析可以看出, 参数回归模型在应用上有一定的局限性, 如果当分析者对参数模型形式的信息了解甚少时, 利用这类模型去分析实际问题有可能会冒较大的风险, 甚至得出不符合实际的结论.

面对上面提到的问题, 那什么样的建模方式才能更好地探索出因变量和自变量之间蕴涵的复杂关系呢? 一个合理的做法是充分依靠数据本身而不是主观假定某种具体的模型形式, “让数据自己说话”, 即在模型 (1.1) 中, 对函数 $f(\cdot)$ 的具体形式不作任何假定或只作一些简单的光滑性要求, 依靠观测数据寻求 $f(\cdot)$ 的特征. 这时, 模型 (1.1) 称为非参数回归模型. 因此, 非参数回归模型尤其适合于当对回归函数的本质特征了解较少时的回归分析问题. 近三十年来借助于计算机计算能力的快速提高, 非参数回归模型 (特别是当自变量为一维的情况) 得到了广泛的研究, 已经有多种光滑方法提出用以估计未知回归函数.

虽然从理论上讲, 各种非参数光滑方法都可以推广到多个回归变量的情形, 但

是随着回归变量个数的增加, 这些方法都面临着所谓的"维数祸根"(curse of dimensionality) 问题. 这是因为非参数函数估计方法本质上讲都是局部估计或局部光滑, 要想使 $m(x)$ 在 x 点得到比较充分的估计, 必须使得 x 的邻域包含有足够多的数据. 但当 x 为多维数据时, 这个条件不易满足, x 的维数越高, 包含一定数量观测点所对应的邻域将会更大以致失去局部特征 (详细讨论可参见 Hastie 和 Tibshirani(1990)), 也就是说要使得光滑方法在高维情况下仍能适用, 则需要的观测点数据会大到难以实现的地步. 为了克服上面提到的"维数祸根"问题, 近年来统计学家与计量经济学家提出了多种多元非参数建模方法.

1.2 常用非参数光滑方法和半参数模型介绍

非参数光滑方法大致可以分为三类, 第一类权函数估计主要是基于局部加权的思想利用周围的点拟合某点处回归函数的值, 常见的最近邻估计、核估计以及后来提出的局部多项式估计都属于该类方法. 第二类最小二乘方法主要是利用参数空间来逼近 (近似) 无穷维参数空间, 基于不同的逼近思想构造参数空间的基函数, 将未知函数 (无穷维参数) 的估计问题转为 (有限个) 未知参数的估计, 从而可以通过最小二乘方法得到未知参数的估计. 该类方法主要包括 Fourier 级数估计、小波方法、多项式 (或分块多项式) 方法以及样条估计. 文献中也常称基于这类方法的估计为 Sieve 估计. 第三类光滑样条估计是一种惩罚最小二乘方法. 关于核估计 (包括局部多项式估计) 方法的详细内容可参考 Muller (1988), Nadaraya (1989), Hardle (1990,1991), Wand 和 Joner (1995), Fan 和 Gijbels (1996), Simonoff (1996), Bowman 和 Azzalini (1997), Hart (1997), Loader (1999), Pagan 和 Ullah (1999) 与 Fox (2000). 关于各类样条方法的详细内容可参考 Eubank (1988), Wahba (1990), Green 和 Silverman (1994), Dierckx (1995), Gu (2002), Ruppert, Wand 和 Carroll (2003) 和 Hansen 等 (2003). 关于经典级数估计的详细内容可参考 Thompson 和 Tapia(1990), Tarter 和 Lock(1993) 和 Efromovich (1999). 关于小波方法的详细内容可参考 Ogden (1996), Louis, Maass 和 Rieder (1997), Hardle 等 (1998), Schimerk (2000) 和 Walter 和 Shen (2001). 下面将简单介绍几类常用的光滑方法以及几种常见的多元非 (半) 参数模型.

考虑如下的一元非参数回归模型

$$Y_i = m(x_i) + \varepsilon_i, \quad i = 1, 2, \cdots, n, \tag{1.4}$$

其中 Y_i 是被解释变量观测值, x_i 为解释变量观测值, x 既可以是随机变量也可以是固定设计变量, 其主要目的是基于观测数据 $\{Y_i, x_i\}_{i=1}^n$ 得到未知回归函数 $m(x)$ 的估计值 $\hat{m}_n(x)$.

1.2.1　权函数估计

假设要估计 x_0 对应的 $m(x_0)$, 一个直观的想法就是利用周围的点, 其中距离 x_0 越近的点对 $m(x_0)$ 的估计所起的作用越大, 距离 x_0 越远的点对 $m(x_0)$ 的估计所起的作用越小. 而这种作用大小的度量主要用权函数来表达. 如果令 $\{y_i, x_i\}$ 在估计 $m(x_0)$ 中对应的权函数为 $W_{ni}(x_0)$, 则有 $m(x_0)$ 的加权估计为

$$\hat{m}(x_0) = \sum_{i=1}^{n} W_{ni}(x_0) Y_i.$$

显然常用的三点平均、分段平均光滑以及最近邻估计都属于简单的权函数估计方法. 下面介绍应用非常广泛的核估计方法.

核估计方法

设 $K(\cdot)$ 是一个实值函数, 通常为对称的概率密度函数并称为核函数, h 是用以控制局部邻域大小的非负实数称为窗宽 (bandwidth) 或者光滑参数 (smoothing parameter). 令 $K_h(\cdot) = K(\cdot/h)/h$, $m(x_0)$ 的 Nadaraya-Watson 核估计为

$$\hat{m}(x_0) = \frac{\sum_{i=1}^{n} K_h(x_i - x_0) Y_i}{\sum_{i=1}^{n} K_h(x_i - x_0)},$$

通常使用的核函数包括 Gauss 核函数

$$K(t) = \frac{1}{\sqrt{2\pi}} \exp(-t^2/2)$$

和对称 β 族函数

$$K(t) = \frac{1}{\beta\left(\frac{1}{2}, \gamma + 1\right)} (1 - t^2)_+^{\gamma},$$

其中“+”表示函数 $(1-t^2)^{\gamma}$ 的正部, $\beta(\cdot,\cdot)$ 为 β 函数. 当 $\gamma = 0, 1, 2, 3$ 时, 分别对应于均匀 (uniform) 核函数、Epanechnikov 核函数、二次 (biweight) 核函数和立方权 (triwight) 函数.

上面估计中随机分母给估计性质的研究带来一定的困难, 为了克服这一缺点, Gasser 和 Muller 提出了一种新的估计方法. 令 $x_0 = -\infty, x_{n+1} = +\infty$, 并假定 $x_0 < x_1 \leqslant x_2 \leqslant \cdots \leqslant x_n < x_{n+1}$, 定义新的估计为

$$\hat{m}(x_0) = \sum_{i=1}^{n} \int_{s_{i-1}}^{s_i} K_h(u - x_0) \mathrm{d}u Y_i,$$

其中 $s_i = \dfrac{x_i + x_{i+1}}{2}$.

虽然核估计在一定的条件下具有相合性与渐近正态性, 但该估计存在边界效应, 即估计在边界处收敛于实际函数的速度慢于在内点处的收敛速度. 而且 Nadaraya-Watson 估计的偏差通常较大, 尤其是在回归函数的导数值较大的地方, 而 Gasser-Muller 估计的方差通常较大.

局部多项式估计方法

注意到上面的核估计可以认为是一种加权最小二乘估计, 即对于 $m(x_0)$ 通过使得

$$\sum_{i=1}^{n}\{Y_i - m(x_0)\}^2 K_h(x_i - x_0)$$

达到最小来估计. 为了克服核估计的缺点, 局部多项式估计方法得到了深入的研究.

设回归函数 $m(x)$ 有 $p+1$ 阶连续导数, 则由 Taylor 展式可知, 对 $m(x)$ 定义域内的 x_0, 其邻域内的函数值 $m(x)$ 有

$$m(x) \approx \sum_{j=0}^{p} \frac{m^{(j)}(x_0)}{j!}(x - x_0)^j \doteq \sum_{j=0}^{p} \beta_j(x_0)(x - x_0)^j,$$

其中 $m^{(j)}(\cdot)$ 表示 $m(\cdot)$ 的第 j 阶导数. 利用局部加权多项式回归方法估计, 对于 $\beta_j(x_0)$, 使得

$$\sum_{i=1}^{n}\{Y_i - \sum_{j=0}^{p}\beta_j(x_0)(x - x_0)^j\}^2 K_h(x_i - x_0)$$

达到最小予以估计. 记 $\beta_j(x_0)$ 的估计为 $\hat{\beta}_j(x_0)$, 则得回归函数及其各阶导数在 x_0 处的估计为

$$\hat{m}^{(v)}(x_0) = v!\hat{\beta}_j(x_0), \quad v = 0, 1, \cdots, p,$$

特别是当上面 $p = 1$ 时, 即将 $m(x)$ 线性展开, 称对应的估计为局部线性估计.

下面对 Nadaraya-Watson 估计、Gasser-Muller 估计以及局部线性估计这三种估计的偏差以及方差做一对比, 内容见下表. 其中记 Bias 和 Var 分别为 $\hat{m}(x_0)$ 的条件偏差与条件方差. 记 $b_n = \dfrac{h^2}{2}\displaystyle\int u^2 K(u)\mathrm{d}u$, $V_n = \dfrac{\sigma^2(x_0)}{nhf(x_0)}\displaystyle\int K^2(u)\mathrm{d}u$.

方法	Bias	Var
Nadaraya-Watson 估计	$\left(m''(x_0) + \dfrac{2m'(x_0)f'(x_0)}{f(x_0)}\right) b_n$	V_n
Gasser-Muller 估计	$m''(x_0)b_n$	$1.5V_n$
局部线性估计	$m''(x_0)b_n$	V_n

从上表可以看到，局部线性估计与 Gasser-Muller 估计相比，二者偏差一样，前者的方差小于后者的方差. 而局部线性估计与 Nadaraya-Watson 估计相比，二者方差一样，前者的偏差小于后者的偏差，而 Nadaraya-Watson 估计与 Gasser-Muller 估计相比，前者的偏差大于后者的偏差，但是方差小于后者的方差. 所以局部线性估计是优于 Gasser-Muller 估计与 Nadaraya-Watson 估计的.

一般来讲局部多项式方法具有以下优点.

(i) 局部多项式估计有相对小的偏差和方差. 与 Nadaraya-Watson 估计以及 Gasser-Muller 估计对应的比较见上表.

(ii) 局部多项式方法适用于各种设计. 它适用于随机设计 (random designs)、固定设计 (fixed designs)、均匀设计 (uniform designs)、分组设计 (clustered designs) 等.

(iii) 局部多项式估计没有边界效应. 局部多项式估计在边界点的估计偏差与内点的估计偏差阶数一样，不需要在边界点处用特殊的权函数来减少边界效应.

(iv) 局部多项式估计有很好的极小极大效率 (minimax efficiency).

1.2.2　样条方法

样条方法可以看成多项式回归的一种推广. 利用分段不同阶数的多项式拟合数据，使得两个多项式函数在连接点 (knots) 处可以允许有不连续的导数，这样便使得估计的回归函数比整体拟合一个固定阶数的多项式更具灵活性. 通常使用的是三次样条，即利用有二阶连续导数的分段多项式函数逼近未知的回归函数. 具体的构造思想如下.

设 $t_1, t_2, \cdots, t_J$ 为固定节点，满足 $-\infty < t_1 < t_2 < \cdots < t_J < +\infty$，即这些节点将实直线划分为如下的区间 $(-\infty, t_1], [t_1, t_2], \cdots, [t_{J-1}, t_J], [t_J, +\infty)$. 三次样条函数 $s(x)$(这里以三次样条为例，其余阶数的情形类似) 有连续二阶导数，并且在上面每个小区间里都是 (不超过) 三次多项式. 所有的三次样条函数形成一个 $J+4$ 维的线性空间，这个空间常用的三次样条基有如下两个：

(1) 幂基 (power basis):　$1, x, x^2, x^3, (x-t_j)_+^3 (j = 1, 2, \cdots, J)$;

(2) B 样条基 (详细的定义以及相关性质可参见 Boor (1978) 和 Schmumaker (1981)).

设选定的三次样条基为 $B_1, \cdots, B_{J+4}$，则三次样条函数为

$$s(x) = \sum_{j=1}^{J+4} \theta_j B_j,$$

上面的未知参数 θ_j 可通过极小化

$$\sum_{i=1}^{n}\left\{Y_i-\sum_{j=1}^{J+4}\theta_jB_j(x_i)\right\}^2$$

予以估计. 记 θ_j 的估计为 $\hat{\theta}_j, j=1,\cdots,J+4$, 从而可以定义 $m(x)$ 的估计为

$$\hat{m}(x)=\sum_{j=1}^{J+4}\hat{\theta}_jB_j(x).$$

其中的 J 称为光滑参数, 其作用相当于前面介绍的局部估计中的窗宽 h. m 次 B 样条函数的基由节点唯一确定, 若选择均匀节点或其他固定方式确定节点的位置, 则基由节点个数唯一确定. 节点个数 k 在拟合数据和回归函数估计的光滑程度之间起平衡作用, 随着节点数目的增加, B 样条估计的方差越大, 偏差越小, 此时该估计更加充分地拟合数据; 随着节点数目的减少，B 样条估计的方差越小, 偏差越大, 估计出的回归函数越光滑. 因此节点控制回归函数的光滑程度.

光滑样条方法是自动选取节点的方法. 如果单纯考虑最小二乘估计, 即求函数 m, 使得

$$\sum_{i=1}^{n}\{Y_i-m(x_i)\}^2$$

达到最小, 显然符合条件的函数 m 为将所有观测点依次连接起来的折线, 这样的估计是没有任何意义的. 为此, 将对 m 附加一定的限制条件, 比如, 要求 m 具有二阶连续导数 (其他阶类似), 那么现在要极小化

$$\sum_{i=1}^{n}\{Y_i-m(x_i)\}^2+\lambda\int(m''(x))^2\mathrm{d}x$$

来估计 m. 上面的解是区间 $[x_{(1)},x_{(n)}]$ 上的唯一的一个三次样条, 且对于任一 x, 其解是 $Y_1,Y_2,\cdots,Y_n$ 的线性函数. 其中 λ 称为光滑参数.

回归 B 样条与惩罚样条和光滑样条都是用样条函数来估计未知的非参数回归函数, 主要区别在于惩罚样条先保守地选择较多数目的节点 (光滑样条甚至选择每一个数据点为节点), 然后再通过惩罚系数防止估计过分地拟合数据, 调节估计函数的光滑程度. 而回归 B 样条估计直接通过节点来平衡拟合数据和估计函数的光滑程度, 所选择的光滑参数与惩罚样条不同, 因此一般回归 B 样条估计所需要的节点个数较少, 从而待估的参数也较少.

1.2.3 部分线性模型

为了克服线性模型的局限性及非参数回归模型可能带来的“维数祸根”问题, 一个努力的方向是 Engle 等 (1986) 提出的部分线性回归模型, 也称为半参数回归

模型. 其一般形式是

$$Y = X_1\beta_1 + X_2\beta_2 + \cdots + X_m\beta_m + f(T) + \varepsilon, \tag{1.5}$$

其中 $(X_1, \cdots, X_p, T)$ 是自变量, 而 Y 为因变量, ε 为模型误差. 可以看出模型 (1.5) 有两部分组成, 线性部分 $X_1\beta_1 + X_2\beta_2 + \cdots + X_m\beta_m$ 称为参数分量, 非线性部分 $g(t)$ 称为非参数分量. 线性部分中的未知参数 β_j 与未知函数 $f(\cdot)$ 的估计及其性质是研究该模型的一个主要任务.

模型 (1.5) 自从 1986 年 Engle 等用于讨论关于气温与供电及电力价格的关系提出以来, 受到了统计学家与计量经济学家的广泛关注, 基本上也可以说是得到研究最多的一类多元非 (半) 参数模型. Heckman (1986), Speckman (1988), Chen (1988), Donaldand 和 Newey (1994), Yatchew (1997) 分别基于不同的光滑思想研究了模型的估计问题; Severini 和 Staniswalis (1994) 研究了广义部分线性模型, 即因变量为离散数据的情况; Liang 等 (1999) 研究了变量含误差 (error in variable) 部分线性模型; 而 Wang 等 (2004) 以及 Liang 等 (2004) 基于该模型分别研究了因变量与自变量存在缺失的情况; Zhang 等 (1998) 利用该建模方法研究了纵向数据 (longitudinal data), Li 和 Ullah (1998) 研究了 Panel data. 关于该模型的详细论述包括一些实际数据的分析可参见柴根象和洪圣岩 (1995), Hardle, Liang 和 Gao (2000), Yatchew (2003), Hardle 等 (2004).

1.2.4　变系数模型

变系数模型的建模思想早已存在, 但直到 Hastie 和 Tibshirani (1993) 的文献发表才引起了人们的关注. 其形式如下

$$Y = \sum_{j=1}^{p} \beta_j(U)X_j + \varepsilon, \tag{1.6}$$

其中 $(U, X_1, \cdots, X_p)$ 是自变量, Y 为因变量, $\beta_j(\cdot)$ 为未知光滑函数, ε 为模型误差.

对于模型中 $\beta_j(\cdot)$ 的估计, Hastie 和 Tibshirani (1993) 提出了利用局部加权核方法以及光滑样条方法拟合模型; Fan 和 Zhang (1999) 基于局部多项式方法提出了两步估计; Eubank 等 (2004) 深入地讨论了光滑样条估计. 由于其良好的解释能力, 该模型在很多领域得到了应用. Cai, Fan 和 Li (2000) 研究了广义变系数模型; Fan, Zhang 和 Zhang (2001) 提出了广义似然比检验, 研究了针对该模型的假设检验问题; Chen 和 Tsay (1993) 以及 Cai, Fan 和 Yao (2000) 将该模型用到了非线性时间序列数据上; Fotheringham, Brundom 和 Charlton (2002) 将该模型用到了空间数据分析上, 用以揭示空间数据的非平稳特征, 并将该模型称之为地理加权回归 (geographically weighted regression) 模型. 在纵向数据的处理上, 该模型可以很好

地揭示各协变量对因变量影响的变化趋势, 从而得到了人们的广泛研究, 针对纵向数据变系数模型的估计、核方法 (Wu 和 Chiang(2000))、局部线性方法 (Hoover et al., 1998)、基函数法 (Huang et al., 2002)、光滑样条 (Chiang et al., 2001)、多项式样条方法 (Huang et al., 2004) 都已经得到了研究. 关于变系数模型的介绍可参考王启华, 史宁中和耿直 (2010).

1.2.5 可加模型

为了使得模型更易解释且不受“维数祸根”问题的影响, Hastie 和 Tibshirani (1986) 提出了可加模型并给出了相应的拟合方法. 该模型为

$$Y=\alpha+f_1(X_1)+f_2(X_2)+\cdots+f_p(X_p)+\varepsilon, \tag{1.7}$$

其中 $X_1,\cdots,X_p$ 为自变量, Y 为因变量, $f_j(\cdot)$ 为任意的一元函数且只与相应的回归变量有关.

对于未知函数 $f_j(\cdot)$, 主要有两种估计方法: 第一种为 Friedman 和 Stuetzle (1981) 提出的 Back-Fitting 算法, 该算法被 Hastie 和 Tibshirani (1986) 用到可加模型以及广义可加模型上, 在 R(S-Plus) 上并有专门的程序包; 第二种方法为 Linton 和 Nielsen (1995), Linton (1997) 提出的边界积分法 (marginal intergation method). Hastie 和 Tibshirani (1990) 是关于该模型早期研究的一个系统总结, 关于最近的研究内容可参考 Hardle 等 (2004).

1.3 预 备 知 识

在非参数与半参数模型的检验中, 经常遇到如下形式的检验统计量

$$T=\frac{\boldsymbol{\varepsilon}^{\mathrm{T}}\boldsymbol{P}_1\boldsymbol{\varepsilon}}{\boldsymbol{\varepsilon}^{\mathrm{T}}\boldsymbol{P}_2\boldsymbol{\varepsilon}}, \tag{1.8}$$

如果令 T 的观测值为 t, 则 T 的检验 p 值为

$$p_0=P_{H_0}(T>t)=P_{H_0}\{\boldsymbol{\varepsilon}^{\mathrm{T}}\boldsymbol{W}\boldsymbol{\varepsilon}>0\}, \tag{1.9}$$

其中 $\boldsymbol{W}=\boldsymbol{P}_1-t\boldsymbol{P}_2$, 即在 H_0 之下，事件 $\{T>t\}$ 发生的概率. 对于给定的显著水平 α，若 $p_0>\alpha$，则接受 H_0；若 $p_0\leqslant\alpha$，则拒绝 H_0. 如果模型误差服从正态分布, 显然上面的检验统计量为正态变量二次型之比的形式. 从而可以利用正态变量二次型分布的有关结果来求得检验 p 值. 为了以后章节的应用, 下面列出一种精确算法和两种逼近算法. 详细内容可参考梅长林 (2000).

1.3.1　精确方法

基于 Imhof (1961) 关于正态变量二次型分布的一个结果, 我们有下面的结论.

结论 1.1　设 n 维随机变量 $\boldsymbol{\xi}=(\xi_1,\xi_2,\cdots,\xi_n)^{\mathrm{T}}\sim N(\boldsymbol{0},\boldsymbol{I})$, $\boldsymbol{A}$为实对称矩阵, 其互不相同的非零特征值为 $\lambda_1,\lambda_2,\cdots,\lambda_m$, 相应的重数分别为 $h_1,h_2,\cdots,h_m$, 则对于二次型 $Q=\boldsymbol{\xi}^{\mathrm{T}}\boldsymbol{A}\boldsymbol{\xi}$, 有

$$P(Q>0)=\frac{1}{2}+\frac{1}{\pi}\int_0^{+\infty}\frac{\sin[\theta(u)]}{u\rho(u)}\mathrm{d}u, \tag{1.10}$$

其中

$$\theta(u)=\frac{1}{2}\sum_{k=1}^{m}[h_k\tan^{-1}(\lambda_k u)], \tag{1.11}$$

$$\rho(u)=\prod_{k=1}^{m}(1+\lambda_k^2u^2)^{h_k/4}, \tag{1.12}$$

进一步, 若令

$$T_U=\left(\pi HU^H\prod_{k=1}^{m}|\lambda_k|^{h_k/2}\right)^{-1}, \tag{1.13}$$

则有 $|t_U|<T_U$，其中 $U>0$ 为任意常数，$H=\dfrac{1}{2}\displaystyle\sum_{k=1}^{m}h_k$.

实际应用中，补充式 (1.10) 中被积函数在原点的定义为 $\lim\limits_{t\to 0}\dfrac{\sin[\theta(t)]}{t\rho(t)}=\dfrac{1}{2}\displaystyle\sum_{k=1}^{m}h_k\lambda_k$，对于任一 $U>0$，可利用数值方法计算积分 $\displaystyle\int_0^U\frac{\sin[\theta(t)]}{t\rho(t)}\mathrm{d}t$, 并可由式 (1.13) 估计其余项. 但用数值方法计算 $\displaystyle\int_0^U\frac{\sin[\theta(t)]}{t\rho(t)}\mathrm{d}t$ 所造成的误差取决于具体的数值积分方法.

显然可以直接利用该推论计算式 (1.9), 将推论中的 $\boldsymbol{A}$ 换作 $\boldsymbol{W}$. 值得注意的是由于 T 关于 $\boldsymbol{\varepsilon}$ 的常数倍具有不变性, 所以假定模型 $\boldsymbol{\varepsilon}\sim N(\boldsymbol{0},\boldsymbol{I})$ 不影响最后结果.

1.3.2　三阶矩 χ^2 逼近方法

前面给出的是精确的计算公式, 计算量非常大. 下面给出计算 p 值的三阶矩逼近方法, 该方法可以大大降低计算量. 三阶矩 χ^2 逼近最早由 Pearson (1959) 用于逼近非中心 χ^2 分布, Imhof (1961) 将其应用于逼近正态变量二次型的分布.

结论 1.2　设式 (1.9) 中的误差项 $\boldsymbol{\varepsilon}$ 为正态分布 $N(0,\boldsymbol{I}_n)$，且 $\mathrm{tr}(\boldsymbol{M})^3\neq 0$. 若应用三阶矩 χ^2 逼近求检验 p 值 p_0，则有

(i) 当 $\mathrm{tr}(\boldsymbol{M}^3) > 0$ 时,

$$p_0 = P_{H_0}(R > r) \approx P(\chi_d^2 > d - h), \tag{1.14}$$

(ii) 当 $\mathrm{tr}(\boldsymbol{M}^3) < 0$ 时,

$$p_0 = P_{H_0}(R > r) \approx P(\chi_d^2 < d - h), \tag{1.15}$$

其中 χ_d^2 是自由度为 d 的 χ^2 变量, 且

$$\begin{cases} d = \dfrac{\{\mathrm{tr}(\boldsymbol{M}^2)\}^3}{\{\mathrm{tr}(\boldsymbol{M}^3)\}^2}, \\ h = \dfrac{\mathrm{tr}(\boldsymbol{M}^2)\ \mathrm{tr}(\boldsymbol{M})}{\mathrm{tr}(\boldsymbol{M}^3)}. \end{cases} \tag{1.16}$$

1.3.3 *F* 分布逼近法

前面给出了计算统计量检验 p 值的精确方法和三阶矩 χ^2 逼近方法, 下面给出一种更简单的方法, 我们称之为 F 分布逼近法, 即用 F 分布逼近统计量 T 的分布, 其主要思想是分别用某适当倍数和自由度的 χ^2 分布逼近 T 中的分子与分母中的正态变量二次型的分布, 然后再用相应自由度的 F 分布逼近 T 的分布, 该方法也是 Cleveland 和 Devlin (1988) 等在讨论非参数回归模型中的假设检验问题中使用的.

结论 1.3 设式 (1.9) 中的误差项 ε 为正态分布 $N(0, \boldsymbol{I}_n)$, 若应用 F 分布逼近方法求检验 p 值 p_0, 则有

$$p = P(T > t) \approx P\left(F(r_1, r_2) > \frac{\mathrm{tr}\boldsymbol{A}_2}{\mathrm{tr}\boldsymbol{A}_1} t\right),$$

其中 $r_1 = \dfrac{[\mathrm{tr}(\boldsymbol{A}_1)]^2}{\mathrm{tr}(\boldsymbol{A}_1^2)}, r_2 = \dfrac{[\mathrm{tr}(\boldsymbol{A}_2)]^2}{\mathrm{tr}(\boldsymbol{A}_2^2)}$.

第 2 章　部分线性变系数模型的 profile Lagrange 乘子检验与 backfitting 估计

部分线性变系数模型作为变系数模型与部分线性模型的推广, 是一种应用非常广泛的数据分析模型. 类似于其他半参数模型, 模型中参数分量 (常值系数) 的估计和检验往往是我们感兴趣的问题. 本章首先研究的是针对常值系数的假设检验问题, 在传统的 Lagrange 乘子检验基础上提出了一种新的 profile Lagrange 乘子检验统计量, 并证明了该统计量在原假设下的渐近分布为卡方分布, 从而将 Lagrange 乘子检验方法推广到半参数模型上. 本章的另外一部分是从可加模型的角度给出了部分线性变系数模型的 backfitting 估计, 并研究了常值系数估计的渐近性质. 本章的主要内容来自于文献魏传华和吴喜之 (2008b, 2008c).

2.1　引　　言

为了更好地探求变量之间蕴涵的复杂关系, 近三十年来借助于计算机强大计算能力而发展起来的非参数回归方法得到了人们的广泛关注. 非参数回归方法克服了参数回归方法主观假定函数形式的严重缺陷, 充分依靠数据本身探求变量之间的关系. 但是非参数回归方法在处理高维数据时会遇到“维数祸根”问题, 为了克服这一局限性, 多种多元非参数和半参数模型相继提出, 比如可加模型 (Breiman and Friedman, 1985; Hastie and Tibshirani, 1990)、变系数模型 (Hastie and Tibshirani, 1993; Fan and Zhang, 1999)、部分线性模型 (Green and Silverman, 1994; Hardle et al., 2000)、多指标模型 (Hardle and Stoker, 1989) 以及这些模型的混合形式 (Carroll et al. (1997)). 其中的变系数模型在最近得到了人们的重视. 该模型的结构如下

$$Y=\sum_{j=1}^{p}\beta_j(U)X_j+\varepsilon, \tag{2.1}$$

其中 $(U,X_1,\cdots,X_p)$ 是自变量, Y 为因变量, ε 为模型误差.

对于变系数模型 (2.1), 实际问题研究中最为普遍的情况是一部分自变量对因变量的影响是随变量 U 而改变的, 但另外一部分自变量对因变量的影响是常值, 此时变系数模型转化为如下的部分线性变系数模型, 其形式可记为

$$Y=\boldsymbol{\alpha}^{\mathrm{T}}(U)\boldsymbol{X}+\boldsymbol{\beta}^{\mathrm{T}}\boldsymbol{Z}+\varepsilon, \tag{2.2}$$

其中 $(U, \boldsymbol{X}^{\mathrm{T}}, \boldsymbol{Z}^{\mathrm{T}})$ 是自变量, 不失一般性, 下面的叙述中假定 U 为一维变量, Y 为因变量, ε 为模型误差, 有 $E(\varepsilon)=0$ 和 $\mathrm{Var}(\varepsilon)=\sigma^2$, 常值系数 $\boldsymbol{\beta}=(\beta_1,\beta_2,\cdots,\beta_q)^{\mathrm{T}}$ 为 q 维未知待估参数, $\boldsymbol{\alpha}(\cdot)=(\alpha_1(\cdot),\alpha_1(\cdot),\cdots,\alpha_p(\cdot))^{\mathrm{T}}$ 为一列未知函数. 一般称 $\boldsymbol{\alpha}^{\mathrm{T}}(U)\boldsymbol{X}$ 为模型的非参数部分, 而称 $\boldsymbol{\beta}^{\mathrm{T}}\boldsymbol{Z}$ 为模型的线性部分. 显然, 当 $\boldsymbol{Z}=\boldsymbol{0}$ 时, 该模型即为形式 (2.1) 的变系数模型. 当 $p=1, \boldsymbol{X}=1$ 时, 模型即为已经得到了充分研究的部分线性模型.

Zhang 等 (2002) 基于局部多项式估计最早研究了模型 (2.2), Zhou 和 You (2004) 基于小波方法研究了该模型, Xia 等 (2004) 基于局部线性方法对该模型提出了一种新的有效估计, Ahmad 等 (2005) 提出了级数估计, Fan 和 Huang (2005) 对参数分量提出了 profile 最小二乘估计方法, 并基于广义似然比方法研究了模型的检验问题.

正如其他的半参数模型, 模型 (2.2) 中常值系数 $\boldsymbol{\beta}$ 往往是研究的重点, 本章首先要研究的是如下的假设检验问题

$$H_0: \boldsymbol{A}\boldsymbol{\beta}=\boldsymbol{b}, \tag{2.3}$$

其中 $\boldsymbol{A}$ 为 $k\times q$ 的已知矩阵, 且 $\mathrm{rank}(\boldsymbol{A})=k$, $\boldsymbol{b}$ 为 $k\times 1$ 已知向量. 同线性模型一样, 模型 (2.2) 中的很多问题都可以归结到形如 (2.3) 的假设检验问题, Fan 和 Huang (2005) 基于广义似然比方法也讨论了这种类型的假设检验问题. 2.2 节将给出模型在原假设条件下的约束 profile 最小二乘估计, 并且基于约束估计构造了 profile Lagrange 乘子 (PLM) 检验统计量.

部分线性变系数模型从结构上可以看成一种特殊的具有两部分的可加模型, 2.3 节将基于 Backfitting 方法给出了模型 (2.2) 的估计, 并研究了常值系数估计的渐近性质.

2.2 profile Lagrange 乘子检验法

最近有多种方法提出用以估计模型 (2.2), 下面将采用 Fan 和 Huang(2005) 提出的基于局部线性光滑的 profile 最小二乘估计方法. 为了叙述的连贯性, 下面首先对该估计方法做一简单介绍.

假设有对模型 (2.2) 的 n 组观测值为 $\{u_k, x_{k1},\cdots,x_{kp},z_{k1},\cdots,z_{kq},y_k\}$, $k=1,2,\cdots,n$. 首先假定模型 (2.2) 中 $(\beta_1,\beta_2,\cdots,\beta_q)^{\mathrm{T}}$ 已知, 则模型转化为如下形式的变系数模型

$$y_i^*=\alpha_1(u_i)x_{i1}+\cdots+\alpha_p(u_i)x_{ip}+\varepsilon_i, \quad i=1,2,\cdots,n, \tag{2.4}$$

其中 $y_i^* = y_i - (z_{i1}\beta_1 + \cdots + z_{iq}\beta_q)$. 针对变系数模型 (2.4), 利用基于局部线性光滑的局部加权最小二乘法来估计未知系数函数. 给定 u_0 邻域内的一点 u, 对 $\alpha_j(u)$ 利用 Taylor 展开式有

$$\alpha_j(u) \approx \alpha_j(u_0) + \alpha_j'(u_0)(u - u_0), \quad j = 1, 2, \cdots, p, \tag{2.5}$$

从而可通过针对 $a_j(u_0), a_j'(u_0)$ 的极小化

$$\sum_{i=1}^{n}\left[y_i^* - \sum_{j=1}^{p}\{\alpha_j(u_0) + \alpha_j'(u_0)(u - u_0)\}x_{ij}\right]^2 K_h(u_i - u_0) \tag{2.6}$$

来估计 $(\hat{\alpha}_j(u_0), \hat{\alpha}_j'(u_0), j = 1, 2, \cdots, p)$, 其中 $K_h(\cdot) = K(\cdot/h)/h$, K 是核函数, h 是窗宽.

为了叙述方便, 采用下面的矩阵形式, 记

$$\boldsymbol{X} = \begin{bmatrix} \boldsymbol{x}_1^{\mathrm{T}} \\ \boldsymbol{x}_2^{\mathrm{T}} \\ \vdots \\ \boldsymbol{x}_n^{\mathrm{T}} \end{bmatrix} = \begin{bmatrix} x_{11} & \cdots & x_{1p} \\ x_{21} & \cdots & x_{2p} \\ \vdots & & \vdots \\ x_{n1} & \cdots & x_{np} \end{bmatrix}, \quad \boldsymbol{Z} = \begin{bmatrix} \boldsymbol{z}_1^{\mathrm{T}} \\ \boldsymbol{z}_2^{\mathrm{T}} \\ \vdots \\ \boldsymbol{z}_n^{\mathrm{T}} \end{bmatrix} = \begin{bmatrix} z_{11} & \cdots & z_{1q} \\ z_{21} & \cdots & z_{2q} \\ \vdots & & \vdots \\ z_{n1} & \cdots & z_{nq} \end{bmatrix},$$

$$\boldsymbol{M} = \begin{bmatrix} \alpha_1(u_1)x_{11} + \cdots + \alpha_p(u_1)x_{1p} \\ \alpha_1(u_2)x_{21} + \cdots + \alpha_p(u_2)x_{2p} \\ \vdots \\ \alpha_1(u_n)x_{n1} + \cdots + \alpha_p(u_n)x_{np} \end{bmatrix} = \begin{bmatrix} \boldsymbol{\alpha}^{\mathrm{T}}(u_1)\boldsymbol{x}_1 \\ \boldsymbol{\alpha}^{\mathrm{T}}(u_2)\boldsymbol{x}_2 \\ \vdots \\ \boldsymbol{\alpha}^{\mathrm{T}}(u_n)\boldsymbol{x}_n \end{bmatrix},$$

$$\boldsymbol{D}_{u_0} = \begin{bmatrix} \boldsymbol{x}_1^{\mathrm{T}} & \boldsymbol{x}_1^{\mathrm{T}}\dfrac{u_1 - u_0}{h} \\ \boldsymbol{x}_2^{\mathrm{T}} & \boldsymbol{x}_2^{\mathrm{T}}\dfrac{u_2 - u_0}{h} \\ \vdots & \vdots \\ \boldsymbol{x}_n^{\mathrm{T}} & \boldsymbol{x}_n^{\mathrm{T}}\dfrac{u_n - u_0}{h} \end{bmatrix},$$

以及 $\boldsymbol{Y} = (y_1, y_2, \cdots, y_n)^{\mathrm{T}}; \boldsymbol{W}_{u_0} = \mathrm{diag}(\mathrm{K}_h(u_1 - u_0), \mathrm{K}_h(u_2 - u_0), \cdots, \mathrm{K}_h(u_n - u_0))$. 从而模型 (2.4) 可写为如下的矩阵形式

$$\boldsymbol{Y} - \boldsymbol{Z\beta} = \boldsymbol{M} + \boldsymbol{\varepsilon}. \tag{2.7}$$

基于问题 (2.6), 由广义最小二乘法可得

$$\left[\hat{\alpha}_1(u_0), \cdots, \hat{\alpha}_p(u_0), h\hat{\alpha}_1'(u_0), \cdots, h\hat{\alpha}_p'(u_0)\right]^{\mathrm{T}}$$

$$=(\boldsymbol{D}_{u_0}^{\mathrm{T}}\boldsymbol{W}_{u_0}\boldsymbol{D}_{u_0})^{-1}\boldsymbol{D}_{u_0}^{\mathrm{T}}\boldsymbol{W}_{u_0}(\boldsymbol{Y}-\boldsymbol{Z}\boldsymbol{\beta}), \tag{2.8}$$

则 $\boldsymbol{M}$ 的初次估计为

$$\widetilde{\boldsymbol{M}}=\boldsymbol{S}(\boldsymbol{Y}-\boldsymbol{Z}\boldsymbol{\beta}),$$

其中

$$\boldsymbol{S}=\begin{bmatrix} (\boldsymbol{x}_1^{\mathrm{T}}\ \boldsymbol{0})\{\boldsymbol{D}_{u_1}^{\mathrm{T}}\boldsymbol{W}_{u_1}\boldsymbol{D}_{u_1}\}^{-1}\boldsymbol{D}_{u_1}^{\mathrm{T}}\boldsymbol{W}_{u_1} \\ (\boldsymbol{x}_2^{\mathrm{T}}\ \boldsymbol{0})\{\boldsymbol{D}_{u_2}^{\mathrm{T}}\boldsymbol{W}_{u_2}\boldsymbol{D}_{u_2}\}^{-1}\boldsymbol{D}_{u_2}^{\mathrm{T}}\boldsymbol{W}_{u_2} \\ \vdots \\ (\boldsymbol{x}_n^{\mathrm{T}}\ \boldsymbol{0})\{\boldsymbol{D}_{u_n}^{\mathrm{T}}\boldsymbol{W}_{u_n}\boldsymbol{D}_{u_n}\}^{-1}\boldsymbol{D}_{u_n}^{\mathrm{T}}\boldsymbol{W}_{u_n} \end{bmatrix}.$$

将 $\boldsymbol{M}$ 的初次估计代入式 (2.7), 整理可得如下的线性回归模型

$$(\boldsymbol{I}-\boldsymbol{S})\boldsymbol{Y}=(\boldsymbol{I}-\boldsymbol{S})\boldsymbol{Z}\boldsymbol{\beta}+\boldsymbol{\varepsilon}. \tag{2.9}$$

利用最小二乘法估计上面的模型, 得 $\boldsymbol{\beta}$ 的 profile 最小二乘估计为

$$\hat{\boldsymbol{\beta}}=[\boldsymbol{Z}^{\mathrm{T}}(\boldsymbol{I}-\boldsymbol{S})^{\mathrm{T}}(\boldsymbol{I}-\boldsymbol{S})\boldsymbol{Z}]^{-1}\boldsymbol{Z}^{\mathrm{T}}(\boldsymbol{I}-\boldsymbol{S})^{\mathrm{T}}(\boldsymbol{I}-\boldsymbol{S})\boldsymbol{Y}, \tag{2.10}$$

得 $\boldsymbol{M}$ 的最终估计为

$$\hat{\boldsymbol{M}}=\boldsymbol{S}(\boldsymbol{Y}-\boldsymbol{Z}\hat{\boldsymbol{\beta}}).$$

从而 $\boldsymbol{Y}$ 的拟合值为

$$\hat{\boldsymbol{Y}}=\hat{\boldsymbol{M}}+\boldsymbol{Z}\hat{\boldsymbol{\beta}}=\boldsymbol{L}\boldsymbol{Y}, \tag{2.11}$$

其中

$$\boldsymbol{L}=\boldsymbol{S}+(\boldsymbol{I}-\boldsymbol{S})\boldsymbol{Z}[\boldsymbol{Z}^{\mathrm{T}}(\boldsymbol{I}-\boldsymbol{S})^{\mathrm{T}}(\boldsymbol{I}-\boldsymbol{S})\boldsymbol{Z}]^{-1}\boldsymbol{Z}^{\mathrm{T}}(\boldsymbol{I}-\boldsymbol{S})^{\mathrm{T}}(\boldsymbol{I}-\boldsymbol{S}).$$

上面拟合过程中, 窗宽 h 可由下述计算量较小的交叉证实法见 (Hastie 和 Tibshirani(1990) 第 3 章) 来确定. 设 $h>0$ 为任一给定的参数值, 为明确拟合值与 h 的关系, 记

$$\hat{\boldsymbol{Y}}(h)=(\hat{y}_1(h),\hat{y}_2(h),\cdots,\hat{y}_n(h))^{\mathrm{T}}=\boldsymbol{L}(h)\boldsymbol{Y},$$

其中 $\boldsymbol{L}(h)=(\boldsymbol{L}_{ij}(h))$ 如式 (2.11) 所示. 令

$$\mathrm{CV}(h)=\sum_{i=1}^{n}\left\{\frac{y_i-\hat{y}_i(h)}{1-\boldsymbol{L}_{ii}(h)}\right\}^2=\sum_{i=1}^{n}\frac{\hat{\varepsilon}_i^2(h)}{[1-\boldsymbol{L}_{ii}(h)]^2}, \tag{2.12}$$

其中 $\hat{\varepsilon}_i(h)=y_i-\hat{y}_i(h)$ $(i=1,2,\cdots,n)$ 为利用所有观测拟合模型 (2.2) 所得残差. 由于 $\boldsymbol{L}$ 有明确的表达式, 因此其主对角线上的各元素 $\boldsymbol{L}_{ii}(h)$ 是容易求得的. 广义交叉证实法即选择 h_0 使得 $\mathrm{CV}(h)$ 达到最小.

约束 profile 最小二乘估计

为了构造检验统计量, 还需要在附加假设条件 (2.3) 的情况下估计模型 (2.2). 由上面的估计过程可以看到 $\boldsymbol{\beta}$ 的估计是基于线性模型 (2.9) 利用最小二乘估计得到的. 下面给出模型的约束 profile 最小二乘估计方法.

基于模型 (2.9) 以及约束条件 (2.3), 构造辅助函数

$$\boldsymbol{F}(\boldsymbol{\beta},\boldsymbol{\lambda})=[(\boldsymbol{I}-\boldsymbol{S})\boldsymbol{Y}-(\boldsymbol{I}-\boldsymbol{S})\boldsymbol{Z}\boldsymbol{\beta}]^{\mathrm{T}}[(\boldsymbol{I}-\boldsymbol{S})\boldsymbol{Y}-(\boldsymbol{I}-\boldsymbol{S})\boldsymbol{Z}\boldsymbol{\beta}]+2\boldsymbol{\lambda}^{\mathrm{T}}(\boldsymbol{A}\boldsymbol{\beta}-\boldsymbol{b}), \quad (2.13)$$

其中 $\boldsymbol{\lambda}$ 为 k 维 Lagrange 乘子. 对函数 $\boldsymbol{F}(\boldsymbol{\beta},\boldsymbol{\lambda})$ 分别对 $\boldsymbol{\beta},\boldsymbol{\lambda}$ 求导, 并令偏导数等于 $\mathbf{0}$, 有

$$\begin{cases}\dfrac{\partial\mathbf{F}(\boldsymbol{\beta},\boldsymbol{\lambda})}{\partial\boldsymbol{\beta}}=-2\boldsymbol{Z}^{\mathrm{T}}(\boldsymbol{I}-\boldsymbol{S})^{\mathrm{T}}(\boldsymbol{I}-\boldsymbol{S})\boldsymbol{Y}+2\boldsymbol{Z}^{\mathrm{T}}(\boldsymbol{I}-\boldsymbol{S})^{\mathrm{T}}(\boldsymbol{I}-\boldsymbol{S})\boldsymbol{Z}\boldsymbol{\beta}+2\boldsymbol{A}^{\mathrm{T}}\boldsymbol{\lambda}=\mathbf{0},\\ \dfrac{\partial\boldsymbol{F}(\boldsymbol{\beta},\boldsymbol{\lambda})}{\partial\boldsymbol{\lambda}}=\boldsymbol{A}\boldsymbol{\beta}-\boldsymbol{b}=\mathbf{0}.\end{cases} \quad (2.14)$$

整理可得 $\boldsymbol{\beta}$ 和 $\boldsymbol{\lambda}$ 的解为

$$\begin{cases}\hat{\boldsymbol{\beta}}_r=\hat{\boldsymbol{\beta}}-(\widetilde{\boldsymbol{Z}}^{\mathrm{T}}\widetilde{\boldsymbol{Z}})^{-1}\boldsymbol{A}^{\mathrm{T}}\left[\boldsymbol{A}(\overline{\boldsymbol{Z}}^{\mathrm{T}}\overline{\boldsymbol{Z}})^{-1}\boldsymbol{A}^{\mathrm{T}}\right]^{-1}(\boldsymbol{A}\hat{\boldsymbol{\beta}}-\boldsymbol{b}),\\ \hat{\boldsymbol{\lambda}}=\left[\boldsymbol{A}(\overline{\boldsymbol{Z}}^{\mathrm{T}}\overline{\boldsymbol{Z}})^{-1}\boldsymbol{A}^{\mathrm{T}}\right]^{-1}(\boldsymbol{A}\hat{\boldsymbol{\beta}}-\boldsymbol{b}),\end{cases} \quad (2.15)$$

其中 $\overline{\boldsymbol{Z}}=(\boldsymbol{I}-\boldsymbol{S})\boldsymbol{Z}$, $\hat{\boldsymbol{\beta}}$ 为 $\boldsymbol{\beta}$ 的无约束 profile 最小二乘估计 (见式 (2.10)). 从而可得到变系数部分 $\boldsymbol{M}$ 的约束估计为

$$\hat{\boldsymbol{M}}_r=\boldsymbol{S}(\boldsymbol{Y}-\boldsymbol{Z}\hat{\boldsymbol{\beta}}_r).$$

最后得 $\boldsymbol{Y}$ 在约束条件下的拟合值为

$$\hat{\boldsymbol{Y}}_r=\hat{\boldsymbol{M}}_r+\boldsymbol{Z}\hat{\boldsymbol{\beta}}_r. \quad (2.16)$$

类似于线性回归模型的约束最小二乘估计, 下面要证明 $\hat{\boldsymbol{\beta}}_r$ 确实是线性约束条件 $\boldsymbol{A}\boldsymbol{\beta}=\boldsymbol{b}$ 下 $\boldsymbol{\beta}$ 的估计, 只需证明如下两点.

(a) $\hat{\boldsymbol{\beta}}_r$ 满足约束条件, 即 $\boldsymbol{A}\hat{\boldsymbol{\beta}}_r=\boldsymbol{b}$.

(b) 对一切满足 $\boldsymbol{A}\boldsymbol{\beta}=\boldsymbol{b}$ 的 $\boldsymbol{\beta}$, 都有 $\|\overline{\boldsymbol{Y}}-\overline{\boldsymbol{Z}}\boldsymbol{\beta}\|^2\geqslant\|\overline{\boldsymbol{Y}}-\overline{\boldsymbol{Z}}\hat{\boldsymbol{\beta}}_r\|^2$.

对于 (a), 由式 (2.15) 即得

$$\boldsymbol{A}\hat{\boldsymbol{\beta}}_r=\boldsymbol{A}\hat{\boldsymbol{\beta}}-(\boldsymbol{A}\hat{\boldsymbol{\beta}}-\boldsymbol{b})=\boldsymbol{b}.$$

对于 (b), 证明方法与线性回归模型约束估计的证明完全相同, 在此省略.

profile Lagrange 乘子检验统计量

针对检验问题 (2.3), Fan 和 Huang (2005) 构造了 profile 广义似然比检验统计量和 Wald 检验统计量, 并给出了这两类统计量在原假设和备择假设下的渐近分布. 下面基于上一部分给出的约束估计, 提出 profile Lagrange 乘子检验统计量并研究其渐近性质.

类似于线性回归模型中 Lagrange 乘子检验统计量的构造, 下面基于 Lagrange 乘子的估计构造检验统计量. 首先假设问题 $H_0: \boldsymbol{A\beta}=\boldsymbol{b}$ 等价于针对 Lagrange 乘子的检验 $H_0^*: \boldsymbol{\lambda}=\boldsymbol{0}$. 由上一部分知 Lagrange 乘子 $\boldsymbol{\lambda}$ 的估计为

$$\hat{\boldsymbol{\lambda}}=\left[\boldsymbol{A}(\overline{\boldsymbol{Z}}^{\mathrm{T}}\overline{\boldsymbol{Z}})^{-1}\boldsymbol{A}^{\mathrm{T}}\right]^{-1}(\boldsymbol{A}\hat{\boldsymbol{\beta}}-\boldsymbol{b}). \tag{2.17}$$

由 $\hat{\boldsymbol{\beta}}$ 的渐近性质 $\hat{\boldsymbol{\beta}}\to N(\boldsymbol{\beta},\sigma^2(\overline{\boldsymbol{Z}}^{\mathrm{T}}\overline{\boldsymbol{Z}})^{-1})$, 得

$$\hat{\boldsymbol{\lambda}}\to N\left(\boldsymbol{A\beta}-\boldsymbol{b},\sigma^2\left[\boldsymbol{A}(\overline{\boldsymbol{Z}}^{\mathrm{T}}\overline{\boldsymbol{Z}})^{-1}\boldsymbol{A}^{\mathrm{T}}\right]^{-1}\right).$$

基于上面的结果, 针对假设检验 H_0^* 构造如下检验统计量

$$T_{\mathrm{PLM}}=\frac{\hat{\boldsymbol{\lambda}}^{\mathrm{T}}\left[\boldsymbol{A}(\overline{\boldsymbol{Z}}^{\mathrm{T}}\overline{\boldsymbol{Z}})^{-1}\boldsymbol{A}^{\mathrm{T}}\right]\hat{\boldsymbol{\lambda}}}{\sigma^2}, \tag{2.18}$$

由于其中的误差方差 σ^2 往往是未知的, 构造其在约束条件下的估计为

$$\hat{\sigma}_r^2=\frac{(\overline{\boldsymbol{Y}}-\overline{\boldsymbol{Z}}\hat{\boldsymbol{\beta}}_r)^{\mathrm{T}}(\overline{\boldsymbol{Y}}-\overline{\boldsymbol{Z}}\hat{\boldsymbol{\beta}}_r)}{n}, \tag{2.19}$$

其中 $\overline{\boldsymbol{Y}}=(\boldsymbol{I}-\boldsymbol{S})\boldsymbol{Y}$. 可以证明 $\hat{\sigma}_r^2$ 在原假设成立时是 σ^2 的相合估计, 所以具体计算中可用 $\hat{\sigma}_r^2$ 代替检验统计量中的 σ^2.

参数回归模型的分析中, 对于类似 (2.3) 形式的线性假设检验问题, 广义似然比检验方法、Wald 检验方法和 Lagrange 乘子检验方法对应的检验统计量在原假设成立的情况下都趋于 χ^2 分布, 即有 Wilks 现象. 对于针对半参数模型 (2.2) 的检验问题 (2.3), 由于讨厌函数 $\boldsymbol{\alpha}(\cdot)$ 是完全非参数化的, 即为无穷维参数, 所以一个自然的问题就是该检验问题对应的统计量在原假设成立的情况下的渐近分布是否为 χ^2 分布? Fan 和 Huang (2005) 证明了 profile 广义似然比检验统计量和 Wald 检验统计量是存在 Wilks 现象的. 下面给出关于前面所构造的 profile Lagrange 乘子检验统计量 T_{PLM} 的回答.

定理 2.1 如果原假设 H_0 成立以及满足 2.5 节的条件 (A.1)~ 条件 (A.6) , T_{PLM} 的渐近分布为 χ_k^2 . 其中 χ_k^2 表示自由度为 k 的 χ^2 分布.

该定理表明 T_{PLM} 的渐近零分布和参数 $\boldsymbol{\beta},\sigma^2$, 以及 U 的设计密度 $f(\cdot)$, 未知系数函数 $\boldsymbol{\alpha}(\cdot)$ 是独立的, 是服从于自由度为 k 的 χ^2 分布. Fan 和 Huang (2005) 提

出的 profile 广义似然比以及 Wald 统计量也具有这些优良的性质, 从而本书的研究也证明了这三个基本的检验方法在处理半参数问题上的渐近等价性, 从而把它们的研究推广到了一个新的领域. 基于上面的结论, 可以利用渐近分布或者模拟求出检验问题 (2.3) 的临界值.

针对原假设 (2.4), 考虑相对应的局部备择假设为

$$H_{\mathrm{a}}: \boldsymbol{A}\boldsymbol{\beta} = \boldsymbol{b} + n^{-1/2}\boldsymbol{\delta}, \tag{2.20}$$

其中 $\boldsymbol{\delta}$ 为一有限常数列.

定理 2.2 在备择假设 H_{a} 情况下, 以及满足式 (2.5) 中的条件 (A.1)~ 条件 (A.6), T_{PLM} 渐近分布为自由度为 k, 非中心参数为 λ 的 χ^2 分布. 其中

$$\lambda = \boldsymbol{\delta}^{\mathrm{T}} \left(\sigma^2 A \boldsymbol{\varXi}^{-1} A^{\mathrm{T}}\right)^{-1} \boldsymbol{\delta} = n(\boldsymbol{A}\boldsymbol{\beta} - \boldsymbol{b})^{\mathrm{T}} \left(\sigma^2 A \boldsymbol{\varXi}^{-1} A^{\mathrm{T}}\right)^{-1} (\boldsymbol{A}\boldsymbol{\beta} - \boldsymbol{b}), \tag{2.21}$$

其中 $\boldsymbol{\varXi} = E(\boldsymbol{Z}\boldsymbol{Z}^{\mathrm{T}}) - E\left[E(\boldsymbol{Z}\boldsymbol{X}^{\mathrm{T}}|U)E(\boldsymbol{X}\boldsymbol{X}^{\mathrm{T}}|U)^{-1}E(\boldsymbol{X}\boldsymbol{Z}^{\mathrm{T}}|U)\right]$.

值得一提的是我们所提出的 profile Lagrange 乘子检验也能很容易地推广到其他类型的半参数模型上, Zhang 等 (2011) 将该方法推广到模型 (2.2) 中自变量 $\boldsymbol{Z}$ 不能精确观测的情形, Feng 和 Xue (2013) 利用该方法研究了模型 (2.2) 中自变量 $\boldsymbol{X}$ 不能精确观测的情形. 利用该检验方法研究单指标模型等复杂形式的半参数模型是值得研究的问题.

2.3 部分线性变系数模型的 backfitting 估计

Backfitting 方法是由 Friedman 和 Stuetzle (1981) 提出用以研究投影寻踪 (projection pursuit) 回归模型的, 后来被 Hastie 和 Tibshirani (1986) 用来拟合可加模型 (additive models), 从而引起了人们的重视. 近年来, 一些文献利用 backfitting 方法研究了其他类型的半参数模型. 对于部分线性模型, Speckmen (1988) 研究了基于核方法的 backfitting 估计. 而 Liang (2006) 研究了基于局部线性方法的 backfitting 估计, Hu 等 (2004) 则基于纵向数据 (longitudinal) 研究了 backfitting 估计, 并且他们都将 backfitting 估计与 profile 最小二乘估计做了比较. 对于部分线性可加模型, Opsomer 和 Ruppert (1999) 研究了线性部分未知参数的 backfitting 估计的渐近性质, 并提供了一种窗宽选择方法.

对于模型 (2.2), 可将其视为由线性部分 $\boldsymbol{\beta}^{\mathrm{T}}\boldsymbol{Z}$ 与变系数部分 $\boldsymbol{\alpha}^{\mathrm{T}}(U)\boldsymbol{X}$ 组成的可加模型, 从而可以利用 backfitting 方法进行估计. 值得注意的是, 如果 U 表示数据 (Y, X, Z) 的 (空间) 观测位置, 那么模型 (2.2) 即为 Fotheringham 等 (2002) 以及魏传华和梅长林 (2005,2006) 研究的半参数空间变系数回归模型 (混合地理加权回

归模型). 魏传华和梅长林 (2006) 结合局部加权回归技术 (基于核方法的局部加权最小二乘估计) 与最小二乘估计给出了模型的 backfitting 估计, 但是该文献只是通过数值实验考察了常值系数 backfitting 估计的优良性, 并没有给出估计的大样本性质. 下面将给出模型 (2.2) 的 backfitting 估计以及常值系数估计的渐近性质, 其中对变系数部分的拟合将基于局部线性方法.

如果模型 (2.2) 中 $(\beta_1, \beta_2, \cdots, \beta_q)^{\mathrm{T}}$ 已知 (给定), 则模型 (2.2) 转化为如下形式的变系数模型

$$y_i^* = \alpha_1(u_i)x_{i1} + \cdots + \alpha_p(u_i)x_{ip} + \varepsilon_i, \quad i = 1, 2, \cdots, n,$$

其中 $y_i^* = y_i - (z_{i1}\beta_1 + \cdots + z_{iq}\beta_q)$. 针对上面的变系数模型, 利用基于局部线性光滑的局部加权最小二乘法来估计未知系数函数. 同 2.2 节中的 profile 最小二乘估计, 可得到变系数部分 $\boldsymbol{M}$ 的拟合值为

$$\hat{\boldsymbol{M}} = \boldsymbol{S}(\boldsymbol{Y} - \boldsymbol{Z}\boldsymbol{\beta}), \tag{2.22}$$

其中 $\boldsymbol{S}$ 同的定义同 2.2 节.

假设模型 (2.2) 中的变系数部分已知, 则模型变为如下的线性回归模型

$$y_i - (\alpha_1(u_i)x_{i1} + \cdots + \alpha_p(u_i)x_{ip}) = z_{i1}\beta_1 + \cdots + z_{iq}\beta_q + \varepsilon_i, \quad i = 1, 2, \cdots, n,$$

从而得 $\boldsymbol{\beta}$ 的最小二乘估计为

$$\hat{\boldsymbol{\beta}} = (\hat{\beta}_1, \cdots, \hat{\beta}_q)^{\mathrm{T}} = [\boldsymbol{Z}^{\mathrm{T}}\boldsymbol{Z}]^{-1}\boldsymbol{Z}^{\mathrm{T}}(\boldsymbol{Y} - \boldsymbol{M}).$$

从而线性部分拟合值为

$$\hat{\boldsymbol{f}} = (\boldsymbol{z}_1^{\mathrm{T}}\hat{\boldsymbol{\beta}}, \boldsymbol{z}_2^{\mathrm{T}}\hat{\boldsymbol{\beta}}, \cdots, \boldsymbol{z}_n^{\mathrm{T}}\hat{\boldsymbol{\beta}})^{\mathrm{T}} = \boldsymbol{Z}\hat{\boldsymbol{\beta}} = \boldsymbol{Z}[\boldsymbol{Z}^{\mathrm{T}}\boldsymbol{Z}]^{-1}\boldsymbol{Z}^{\mathrm{T}}(\boldsymbol{Y} - \boldsymbol{M}). \tag{2.23}$$

根据 backfitting 原理 (详见 Hastie 和 Tibshirani(1990)[118-120]), 由式 (2.22) 和式 (2.23) 得到下面的估计方程

$$\begin{cases} \boldsymbol{M} = \boldsymbol{S}(\boldsymbol{Y} - \boldsymbol{Z}\boldsymbol{\beta}), \\ \boldsymbol{Z}\boldsymbol{\beta} = \boldsymbol{Z}[\boldsymbol{Z}^{\mathrm{T}}\boldsymbol{Z}]^{-1}\boldsymbol{Z}^{\mathrm{T}}(\boldsymbol{Y} - \boldsymbol{M}). \end{cases} \tag{2.24}$$

将第一式代入第二式, 并假定矩阵 $\boldsymbol{Z}^{\mathrm{T}}(\boldsymbol{I} - \boldsymbol{S})\boldsymbol{Z}$ 可逆, 则可求得

$$\begin{cases} \hat{\boldsymbol{\beta}}_{\mathrm{BF}} = [\boldsymbol{Z}^{\mathrm{T}}(\boldsymbol{I} - \boldsymbol{S})\boldsymbol{Z}]^{-1}\boldsymbol{Z}^{\mathrm{T}}(\boldsymbol{I} - \boldsymbol{S})\boldsymbol{Y}, \\ \hat{\boldsymbol{M}}_{\mathrm{BF}} = \boldsymbol{S}(\boldsymbol{Y} - \boldsymbol{Z}\hat{\boldsymbol{\beta}}). \end{cases} \tag{2.25}$$

从而 $\boldsymbol{Y}$ 的拟合值为

$$\hat{\boldsymbol{Y}} = \hat{\boldsymbol{M}}_{\mathrm{BF}} + \boldsymbol{Z}\hat{\boldsymbol{\beta}}_{\mathrm{BF}} = \boldsymbol{L}_{\mathrm{BF}}\boldsymbol{Y},$$

其中

$$\boldsymbol{L}_{\mathrm{BF}} = \boldsymbol{S} + (\boldsymbol{I} - \boldsymbol{S})\boldsymbol{Z}[\boldsymbol{Z}^{\mathrm{T}}(\boldsymbol{I} - \boldsymbol{S})\boldsymbol{Z}]^{-1}\boldsymbol{Z}^{\mathrm{T}}(\boldsymbol{I} - \boldsymbol{S}).$$

容易看出参数分量 $\boldsymbol{\beta}$ 与系数函数 $\boldsymbol{\alpha}(\cdot)$ 的 backfitting 估计与 2.2 节中对应的 profile 最小二乘估计在表达上非常类似. 如果矩阵 $(\boldsymbol{I} - \boldsymbol{S})$ 对称幂等, 则这两种估计方法是相同的. 但是模型的特点决定了矩阵 $(\boldsymbol{I} - \boldsymbol{S})$ 一般不是对称幂等的, 所以这两种估计方法是不同的, 同时也决定了这两种估计在很多方面的差别. 对于部分可加模型以及部分线性模型, 前面提到的 Speckman (1988), Opsomer 和 Ruppert (1999) 和 Liang (2006) 都分别针对常值系数的两种估计做了渐近性质方面的比较.

下面是关于常值系数 $\boldsymbol{\beta}$ 的 Basck-Fitting 估计的性质, 证明过程中需要的一些常规条件将在附录中给出, 另外记 $\mu_i = \displaystyle\int \mu^i K(\mu)\mathrm{d}\mu, \nu_i = \int \mu^i K^2(\mu)\mathrm{d}\mu$, 并令 $\varGamma(U) = E(\boldsymbol{X}\boldsymbol{X}^{\mathrm{T}}|U), \varPhi(U) = E(\boldsymbol{X}\boldsymbol{Z}^{\mathrm{T}}|U)$.

定理 2.3 对于常值系数 $\boldsymbol{\beta}$ 的 backfitting 估计有

$$E(\hat{\boldsymbol{\beta}}_{\mathrm{BF}} - \boldsymbol{\beta}|\boldsymbol{X}, \boldsymbol{Z}, \boldsymbol{U}) = [\boldsymbol{Z}^{\mathrm{T}}(\boldsymbol{I} - \boldsymbol{S})\boldsymbol{Z}]^{-1}\boldsymbol{Z}^{\mathrm{T}}(\boldsymbol{I} - \boldsymbol{S})\boldsymbol{M},$$

$$\mathrm{Var}(\hat{\boldsymbol{\beta}}_{\mathrm{BF}}|\boldsymbol{X}, \boldsymbol{Z}, \boldsymbol{U}) = \sigma^2[\boldsymbol{Z}^{\mathrm{T}}(\boldsymbol{I} - \boldsymbol{S})\boldsymbol{Z}]^{-1}\boldsymbol{Z}^{\mathrm{T}}(\boldsymbol{I} - \boldsymbol{S})(\boldsymbol{I} - \boldsymbol{S})^{\mathrm{T}}\boldsymbol{Z}[\boldsymbol{Z}^{\mathrm{T}}(\boldsymbol{I} - \boldsymbol{S})\boldsymbol{Z}]^{-1}.$$

如果满足 2.5 节的条件 (A.1)~ 条件 (A.6), 则其渐近偏与渐近方差分别为

$$E(\hat{\boldsymbol{\beta}}_{\mathrm{BF}} - \boldsymbol{\beta}|\boldsymbol{X}, \boldsymbol{Z}, \boldsymbol{U}) = -\frac{h^2}{2}\mu_2(K)E[\boldsymbol{z}_i\boldsymbol{x}_i^{\mathrm{T}}\boldsymbol{\alpha}''(u_i)]\boldsymbol{\varXi}^{-1} + o_p(h^2),$$

$$\mathrm{Var}(\hat{\boldsymbol{\beta}}_{\mathrm{BF}}|\boldsymbol{X}, \boldsymbol{Z}, \boldsymbol{U}) = \frac{\sigma^2}{n}\boldsymbol{\varXi}^{-1} + o_p\left(\frac{1}{n}\right),$$

其中 $\boldsymbol{\varXi} = E(\boldsymbol{Z}\boldsymbol{Z}^{\mathrm{T}}) - E\left[E(\boldsymbol{Z}\boldsymbol{X}^{\mathrm{T}}|U)E(\boldsymbol{X}\boldsymbol{X}^{\mathrm{T}}|U)^{-1}E(\boldsymbol{X}\boldsymbol{Z}^{\mathrm{T}}|U)\right]$.

上面的结论表明如果选择一般的窗宽 $h \sim n^{1/5}$, 那么常值系数 $\boldsymbol{\beta}$ 的估计便不是 $\sqrt{n}$ 相合的. 与此不同的是, $\boldsymbol{\beta}$ 的 profile 最小二乘估计在该情况下是 $\sqrt{n}$ 相合的, 这表明要想使得 $\hat{\boldsymbol{\beta}}_{\mathrm{BF}}$ 达到 $\sqrt{n}$ 相合, 拟合不足 (undersoothing) 是不可避免的, 而对于 profile 最小二乘估计, 利用最优窗宽 $h \sim n^{-1/5}$ 即可满足 $\sqrt{n}$ 相合. Spenkman (1988), Liang (2006) 与 Opsomer 和 Ruppert (1999) 对于部分线性模型与部分线性可加模型也得到了同样的结论. 同样地, 如果对窗宽进行调整, 可得到下面的结果.

推论 2.1 如果 $h \sim n^{r/5}$, 其中 $-1 < r < -1/4$, 则有

$$\sqrt{n}(\hat{\boldsymbol{\beta}}_{\mathrm{BF}} - \boldsymbol{\beta}) \to N(0, \sigma^2\boldsymbol{\varSigma}^{-1}).$$

由推论可知, 如果正确的选择窗宽, backfitting 估计一样可以达到半参数效率界限.

2.4 数值模拟

下面将通过数值模拟考察前面所提出的 backfitting 估计的有效性. 下面的模拟实验中核函数选取的是 Epanechnikov 核 $K(u) = 0.75(1-u^2)_+$.

考虑如下的部分线性变系数模型

$$y_i = \beta_0 + \beta_1 x_{i1} + \beta_2(u_i)x_{i2} + \varepsilon_i, \quad i = 1, 2, \cdots, n. \tag{2.26}$$

令 $\beta_0 = 3, \beta_1 = 2, \beta_2(u_i) = \sin(2\pi u_i)$, 自变量 x_{1i} 是独立产生的服从 $U(0,1)$ 分布的随机数, 而 x_{2i} 是服从 $N(2,1)$ 的随机数, 模型误差 ε_i 是来自于 $N(0,\sigma^2)$ 的随机数. 为了考察样本量以及误差方差对估计的影响, 下面分别考察样本量 $n = 100, 200$ 和误差方差 $\sigma^2 = 0.2^2, 0.5^2, 0.8^2$ 几种情况. 每种情况下模拟的重复次数为 500 次, 其中窗宽利用 Cross-Validation 方法进行选择. 下面将基于 500 次对应的均值 (mean) 与标准差 (SD) 来考察估计的有效性, 具体的结果见表 2.1. 从上面的结果可以看出如下结论.

(1) 随着 n 的增大, 即观测点的增多, 虽然估计的精度 (估计值逼近精确值的程度) 提高不大, 但其稳定性有明显的提高.

(2) 随着 σ 的增大, 即噪声方差变大, 对模型的干扰增强, 估计的精度变化不大, 但稳定性有明显的降低.

表 2.1 常值系数的 backfitting 估计的均值与标准差

		$\beta_0 = 3$			$\beta_1 = 2$		
		$\sigma = 0.2$	$\sigma = 0.5$	$\sigma = 0.8$	$\sigma = 0.2$	$\sigma = 0.5$	$\sigma = 0.8$
$n = 100$	均值	2.9868	2.7473	3.6684	2.1762	2.1598	1.6129
	SD	0.0558	0.1432	0.1948	0.0673	0.1779	0.2858
$n = 200$	均值	3.0026	3.2567	3.0708	2.2049	1.6836	1.9189
	SD	0.0427	0.1023	0.1574	0.0466	0.1223	0.1964

下面考察光滑参数对常值系数的 backfitting 估计的影响, 在 $n = 100, \sigma^2 = 0.2^2$ (下面若不做特殊说明, 样本量与误差方差取值与此相同) 的情况下考察光滑参数 h 分别等于 0.1, 0.3, 0.5, 0.7, 0.9 和 1.1 六种情形. 模拟结果见表 2.2. 模拟结果表示窗宽的变化对常值系数的 backfitting 估计的精度以及稳定性影响不大, 这也与关于部分线性模型等半参数模型的研究结果相一致.

表 2.2 不同窗宽下各常值系数 backfitting 估计的均值与标准差

h	0.1	0.3	0.5	0.7	0.9	1.1
Mean(β_0)	3.0152	2.9610	2.9738	2.9495	2.9719	2.999
SD(β_0)	0.0572	0.0628	0.0584	0.0620	0.0584	0.0536
Mean(β_1)	2.0047	2.1071	2.1668	2.1891	2.1806	2.1732
SD(β_1)	0.0048	0.0720	0.0664	0.0699	0.0708	0.0668

接下来将基于估计均方误差$\mathrm{MSE}(\hat{\beta}) = \dfrac{1}{500}\sum_{j=1}^{500}(\hat{\beta}_j - \beta)^2$ 来比较常值系数的 backfitting 估计 (简称 BF 估计) 与 profile 最小二乘估计 (简称 PL 估计) 的有效性, 模拟结果见表 2.3. 模拟结果表明 backfitting 估计与 profile 最小二乘估计在对应的几种窗宽下相差不大. 这也与我们的结果相一致.

表 2.3 不同窗宽下各常值系数 BF 估计与 PL 估计的估计均方误差

h	0.1	0.3	0.5	0.7	0.9	1.1
$\mathrm{MSE_{PL}}(\hat{\beta}_0)$	0.0032	0.0039	0.0059	0.0069	0.0071	0.0088
$\mathrm{MSE_{BF}}(\hat{\beta}_0)$	0.0035	0.0054	0.0040	0.0063	0.0042	0.0028
$\mathrm{MSE_{PL}}(\hat{\beta}_1)$	0.0048	0.0083	0.0171	0.0250	0.0253	0.0217
$\mathrm{MSE_{BF}}(\hat{\beta}_1)$	0.0047	0.0166	0.0322	0.0406	0.0376	0.0344

2.5 定理的证明

下面将给出定理 2.1 至定理 2.3 的证明. 在给出具体证明之前, 首先列出一些条件.

(A.1) 随机变量 U 具有有界支撑 $\boldsymbol{\Omega}$, 其密度函数 $f(\cdot)$ 在其支撑上满足 Lipschitz 连续, 且不为 0.

(A.2) 对于任一 $U \in \boldsymbol{\Omega}$, 矩阵 $E(\boldsymbol{X}\boldsymbol{X}^{\mathrm{T}}|U)_{k\times K}$ 为非奇异, $E(\boldsymbol{X}\boldsymbol{X}^{\mathrm{T}}|U)_{k\times K}$, $E(\boldsymbol{X}\boldsymbol{X}^{\mathrm{T}}|U)_{k\times K}^{-1}$ 和 $E(\mathbf{X}\mathbf{Z}^{\mathrm{T}}|\mathbf{U})_{\mathbf{k}\times\mathbf{K}}$ 都是 Lipschitz 连续的.

(A.3) 存在 $s>2$, 使得 $E\|\boldsymbol{X}\|^{2s}<\infty$ 和 $E\|\boldsymbol{Z}\|^{2s}<\infty$, 对于 $\varepsilon < 2 - s^{-1}$, 使得 $n^{2\varepsilon-1}h \to \infty$.

(A.4) $\alpha(\cdot)$ 二阶连续可导.

(A.5) 函数 $K(\cdot)$ 为对称密度函数, 具有紧支撑.

(A.6) $nh^8 \to 0$ 和 $nh^2/(\log n)^2 \to \infty$.

上述条件是 Fan 和 Huang (2005) 证明定理所附加的条件, 这些条件只是为了证明的方便, 并非最弱条件.

另外为了定理的证明, 给出如下引理.

引理 2.1 令 $(\boldsymbol{X}_1, \boldsymbol{Y}_1), \cdots, (\boldsymbol{X}_n, \boldsymbol{Y}_n)$ 为独立同分布 (iid) 的随机序列, 其中

$\boldsymbol{Y}_i, i=1,2,\cdots,n$ 为一元随机变量, 进一步假定 $E|y|^s<\infty$ 与 $\sup_x\int|y|^s f(x,y)\mathrm{d}y<\infty$, 其中 f 表示 $(\boldsymbol{X},\boldsymbol{Y})$ 的联合密度. 令 K 为一有界正函数, 并有有界支撑且满足 Lipschitz 条件, 则有

$$\sup_x\left|\frac{1}{n}\sum_{i=1}^{n}[K_h(\boldsymbol{X}_i-x)\boldsymbol{Y}_i-E(K_h(\boldsymbol{X}_i-x)\boldsymbol{Y}_i)]\right|=O_p\left(\left\{\frac{\log(1/h)}{nh}\right\}^{1/2}\right), \tag{2.27}$$

其中对于 $\varepsilon<1-s^{-1}$, 有 $n^{2\varepsilon-1}h\to\infty$.

引理 2.1 由 Mack 和 Silverman(1982) 即可证得. 下面记 $c_n=\left\{\dfrac{\log(1/h)}{nh}\right\}^{1/2}$

引理 2.2 如果满足上面的条件 (A.1)~ 条件 (A.6), 那么 $\boldsymbol{\beta}$ 无约束下的 Profiel 最小二乘估计是渐近正态的, 即

$$\sqrt{n}(\hat{\boldsymbol{\beta}}-\boldsymbol{\beta})\to N(\mathbf{0},\sigma^2\boldsymbol{\Xi}^{-1}).$$

引理 2.3 如果满足上面的条件 (A.1)~ 条件 (A.6), 有

$$n^{-1}\overline{\boldsymbol{Z}}^{\mathrm{T}}\overline{\boldsymbol{Z}}\to\boldsymbol{\Xi}.$$

引理 2.4 如果满足上面的条件 (A.1)~ 条件 (A.6), 有

$$\frac{1}{n}\overline{\boldsymbol{Z}}^{\mathrm{T}}(\boldsymbol{I}-\boldsymbol{S})(\boldsymbol{I}-\boldsymbol{S})^{\mathrm{T}}\overline{\boldsymbol{Z}}\to\boldsymbol{\Xi}.$$

注 引理 2.2 至引理 2.4 分别为 Fan 和 Huang (2005) 的定理 4.1, 引理 7.2 与引理 7.3.

定理 2.1 和 2.2 的证明 由引理 2.2 和 2.3 可得, 原假设成立的情况下有

$$\boldsymbol{A}\hat{\boldsymbol{\beta}}-\boldsymbol{b}\to N(\mathbf{0},\hat{\sigma}_r^2\boldsymbol{A}(\overline{\boldsymbol{Z}}^{\mathrm{T}}\overline{\boldsymbol{Z}})^{-1}\boldsymbol{A}^{\mathrm{T}}),$$

从而有

$$\hat{\boldsymbol{\lambda}}\to N(\mathbf{0},\hat{\sigma}_r^2[\boldsymbol{A}(\overline{\boldsymbol{Z}}^{\mathrm{T}}\overline{\boldsymbol{Z}})^{-1}\boldsymbol{A}^{\mathrm{T}}]^{-1}),$$

则有正态分布的性质有

$$\frac{\hat{\boldsymbol{\lambda}}^{\mathrm{T}}\left[\boldsymbol{A}(\overline{\boldsymbol{Z}}^{\mathrm{T}}\overline{\boldsymbol{Z}})^{-1}\boldsymbol{A}^{\mathrm{T}}\right]\hat{\boldsymbol{\lambda}}}{\hat{\sigma}_r^2}\to\chi_k^2.$$

备择假设下, 则有

$$\boldsymbol{A}\hat{\boldsymbol{\beta}}-\boldsymbol{b}\to N(\boldsymbol{A}\boldsymbol{\beta}-\boldsymbol{b},\hat{\sigma}_r^2\boldsymbol{A}(\overline{\boldsymbol{Z}}^{\mathrm{T}}\overline{\boldsymbol{Z}})^{-1}\boldsymbol{A}^{\mathrm{T}}),$$

所以

$$\hat{\boldsymbol{\lambda}} \to N([\boldsymbol{A}(\overline{\boldsymbol{Z}}^{\mathrm{T}}\overline{\boldsymbol{Z}})^{-1}\boldsymbol{A}^{\mathrm{T}}]^{-1}(\boldsymbol{A}\boldsymbol{\beta}-\boldsymbol{b}), \hat{\sigma}_r^2[\boldsymbol{A}(\overline{\boldsymbol{Z}}^{\mathrm{T}}\overline{\boldsymbol{Z}})^{-1}\boldsymbol{A}^{\mathrm{T}}]^{-1}),$$

则有

$$\frac{\hat{\boldsymbol{\lambda}}^{\mathrm{T}}\left[\boldsymbol{A}(\overline{\boldsymbol{Z}}^{\mathrm{T}}\overline{\boldsymbol{Z}})^{-1}\boldsymbol{A}^{\mathrm{T}}\right]\hat{\boldsymbol{\lambda}}}{\hat{\sigma}_r^2} \to \chi^2(k,\lambda),$$

其中 $\chi^2(k,\lambda)$ 为自由度为 k, 非中心参数为 λ 的 χ^2 分布, 非中心参数为

$$\lambda = n(\boldsymbol{A}\boldsymbol{\beta}-\boldsymbol{b})^{\mathrm{T}}\left(\sigma^2 A\boldsymbol{\varXi}^{-1}A^{\mathrm{T}}\right)^{-1}(\boldsymbol{A}\boldsymbol{\beta}-\boldsymbol{b}).$$

证毕.

定理 2.3 的证明 首先, 由

$$\begin{aligned}\hat{\boldsymbol{\beta}}_{\mathrm{BF}} &= [\boldsymbol{Z}^{\mathrm{T}}(\boldsymbol{I}-\boldsymbol{S})\boldsymbol{Z}]^{-1}\boldsymbol{Z}^{\mathrm{T}}(\boldsymbol{I}-\boldsymbol{S})\boldsymbol{Y}\\ &= [\boldsymbol{Z}^{\mathrm{T}}(\boldsymbol{I}-\boldsymbol{S})\boldsymbol{Z}]^{-1}\boldsymbol{Z}^{\mathrm{T}}(\boldsymbol{I}-\boldsymbol{S})(\boldsymbol{Z}\boldsymbol{\beta}+\boldsymbol{M}+\boldsymbol{\varepsilon})\\ &= \boldsymbol{\beta}+[\boldsymbol{Z}^{\mathrm{T}}(\boldsymbol{I}-\boldsymbol{S})\boldsymbol{Z}]^{-1}\boldsymbol{Z}^{\mathrm{T}}(\boldsymbol{I}-\boldsymbol{S})\boldsymbol{M}+[\boldsymbol{Z}^{\mathrm{T}}(\boldsymbol{I}-\boldsymbol{S})\boldsymbol{Z}]^{-1}\boldsymbol{Z}^{\mathrm{T}}(\boldsymbol{I}-\boldsymbol{S})\boldsymbol{\varepsilon},\end{aligned}$$

立即可得

$$E(\hat{\boldsymbol{\beta}}_{\mathrm{BF}}-\boldsymbol{\beta}|\boldsymbol{X},Z,U) = [\boldsymbol{Z}^{\mathrm{T}}(\boldsymbol{I}-\boldsymbol{S})\boldsymbol{Z}]^{-1}\boldsymbol{Z}^{\mathrm{T}}(\boldsymbol{I}-\boldsymbol{S})\boldsymbol{M}. \tag{2.28}$$

为了得到上面偏差的渐近性质, 下面分别讨论 $\boldsymbol{Z}^{\mathrm{T}}(\boldsymbol{I}-\boldsymbol{S})\boldsymbol{Z}$ 与 $\boldsymbol{Z}^{\mathrm{T}}(\boldsymbol{I}-\boldsymbol{S})\boldsymbol{M}$ 这两部分.

对于 $\boldsymbol{Z}^{\mathrm{T}}(\boldsymbol{I}-\boldsymbol{S})\boldsymbol{Z}$, 有

$$\boldsymbol{S}\boldsymbol{Z} = \begin{pmatrix}(\boldsymbol{x}_1^{\mathrm{T}}\ \boldsymbol{0})\{\boldsymbol{D}_{u_1}^{\mathrm{T}}\boldsymbol{W}_{u_1}\boldsymbol{D}_{u_1}\}^{-1}\boldsymbol{D}_{u_1}^{\mathrm{T}}\boldsymbol{W}_{u_1}\boldsymbol{Z}\\ (\boldsymbol{x}_2^{\mathrm{T}}\ \boldsymbol{0})\{\boldsymbol{D}_{u_2}^{\mathrm{T}}\boldsymbol{W}_{u_2}\boldsymbol{D}_{u_2}\}^{-1}\boldsymbol{D}_{u_2}^{\mathrm{T}}\boldsymbol{W}_{u_2}\boldsymbol{Z}\\ \vdots\\ (\boldsymbol{x}_n^{\mathrm{T}}\ \boldsymbol{0})\{\boldsymbol{D}_{u_n}^{\mathrm{T}}\boldsymbol{W}_{u_n}\boldsymbol{D}_{u_n}\}^{-1}\boldsymbol{D}_{u_n}^{\mathrm{T}}\boldsymbol{W}_{u_n}\boldsymbol{Z}\end{pmatrix},$$

其中

$$\boldsymbol{D}_u^{\mathrm{T}}\boldsymbol{W}_u\boldsymbol{D}_u = \begin{pmatrix}\sum_{i=1}^n \boldsymbol{x}_i\boldsymbol{x}_i^{\mathrm{T}}K_h(u_i-u) & \sum_{i=1}^n \boldsymbol{x}_i\boldsymbol{x}_i^{\mathrm{T}}\left(\frac{u_i-u}{h}\right)K_h(u_i-u)\\ \sum_{i=1}^n \boldsymbol{x}_i\boldsymbol{x}_i^{\mathrm{T}}\left(\frac{u_i-u}{h}\right)K_h(u_i-u) & \sum_{i=1}^n \boldsymbol{x}_i\boldsymbol{x}_i^{\mathrm{T}}\left(\frac{u_i-u}{h}\right)^2K_h(u_i-u)\end{pmatrix},$$

$$\boldsymbol{D}_u^{\mathrm{T}}\boldsymbol{W}_u\boldsymbol{Z} = \begin{pmatrix}\sum_{i=1}^n \boldsymbol{x}_i\mathbf{z}_i^{\mathrm{T}}K_h(u_i-u)\\ \sum_{i=1}^n \boldsymbol{x}_i\mathbf{z}_i^{\mathrm{T}}\left(\frac{u_i-u}{h}\right)K_h(u_i-u)\end{pmatrix}.$$

经过类似于密度估计的计算, 由引理 2.1 得

$$\boldsymbol{D}_u^{\mathrm{T}}\boldsymbol{W}_u\boldsymbol{D}_u = nf(u)\boldsymbol{\Gamma}(u)\otimes\begin{pmatrix}1 & 0\\ 0 & \mu_2\end{pmatrix}\{1+O_p(c_n)\}, \tag{2.29}$$

其中 $\otimes$ 表示 Kronecker 积. 同样有

$$\boldsymbol{D}_u^{\mathrm{T}}\boldsymbol{W}_u\boldsymbol{Z} = nf(u)\boldsymbol{\Phi}(u)\otimes(1,0)^{\mathrm{T}}\{1+O_p(c_n)\}. \tag{2.30}$$

所以有

$$\boldsymbol{SZ} = \begin{pmatrix}\boldsymbol{x}_1^{\mathrm{T}}\Gamma^{-1}(u_1)\Phi(u_1)\\ \boldsymbol{x}_2^{\mathrm{T}}\Gamma^{-1}(u_2)\Phi(u_2)\\ \vdots\\ \boldsymbol{x}_n^{\mathrm{T}}\Gamma^{-1}(u_n)\Phi(u_n)\end{pmatrix}\{1+O_p(c_n)\}. \tag{2.31}$$

从而有

$$\frac{1}{n}\boldsymbol{Z}^{\mathrm{T}}(\boldsymbol{I}-\boldsymbol{S})\boldsymbol{Z} = \frac{1}{n}\sum_{i=1}^{n}\mathbf{z}_i\mathbf{z}_i^{\mathrm{T}} - \frac{1}{n}\sum_{i=1}^{n}\mathbf{z}_i\boldsymbol{x}_i^{\mathrm{T}}\Gamma^{-1}(u_i)\Phi(u_i)\{1+O_p(c_n)\}\to\boldsymbol{\Xi}. \tag{2.32}$$

下面讨论 $\boldsymbol{Z}^{\mathrm{T}}(\boldsymbol{I}-\boldsymbol{S})\boldsymbol{M}$, 首先对于系数函数 $\boldsymbol{\alpha}(\cdot)$ 利用 Taylor 展开式, 对 u_0 邻域内的点有

$$\boldsymbol{\alpha}(u_i) = \boldsymbol{\alpha}(u_0) + h\boldsymbol{\alpha}'(u_0)\left(\frac{u_i-u_0}{h}\right) + \frac{h^2}{2}\boldsymbol{\alpha}''(u_0)\left(\frac{u_i-u_0}{h}\right)^2 + o_p(h^2).$$

所以有

$$\begin{aligned}\boldsymbol{M} &= \begin{pmatrix}\boldsymbol{x}_1^{\mathrm{T}}\boldsymbol{\alpha}(u_1)\\ \boldsymbol{x}_2^{\mathrm{T}}\boldsymbol{\alpha}(u_2)\\ \vdots\\ \boldsymbol{x}_n^{\mathrm{T}}\boldsymbol{\alpha}(u_n)\end{pmatrix}\\ &= \begin{pmatrix}\boldsymbol{x}_1^{\mathrm{T}}\{\boldsymbol{\alpha}(u_0) + h\boldsymbol{\alpha}'(u_0)\left(\frac{u_1-u_0}{h}\right) + \frac{h^2}{2}\boldsymbol{\alpha}''(u_0)\left(\frac{u_1-u_0}{h}\right)^2\}\\ \boldsymbol{x}_2^{\mathrm{T}}\boldsymbol{\alpha}(u_0) + h\boldsymbol{\alpha}''(u_0)\left(\frac{u_2-u_0}{h}\right) + \frac{h^2}{2}\boldsymbol{\alpha}'(u_0)\left(\frac{u_2-u_0}{h}\right)^2\\ \vdots\\ \boldsymbol{x}_n^{\mathrm{T}}\boldsymbol{\alpha}(u_0) + h\boldsymbol{\alpha}'(u_0)\left(\frac{u_n-u_0}{h}\right) + \frac{h^2}{2}\boldsymbol{\alpha}''(u_0)\left(\frac{u_n-u_0}{h}\right)^2\end{pmatrix} + o_p(h^2)\end{aligned}$$

$$
= \boldsymbol{D}_{u_0} \begin{pmatrix} \boldsymbol{\alpha}(u_0) \\ h\boldsymbol{\alpha}'(u_0) \end{pmatrix} + \frac{h^2}{2} \begin{pmatrix} \boldsymbol{x}_1^{\mathrm{T}} \left(\dfrac{u_1 - u_0}{h} \right)^2 \\ \boldsymbol{x}_2^{\mathrm{T}} \left(\dfrac{u_2 - u_0}{h} \right)^2 \\ \vdots \\ \boldsymbol{x}_n^{\mathrm{T}} \left(\dfrac{u_n - u_0}{h} \right)^2 \end{pmatrix} \boldsymbol{\alpha}''(u_0) + o_p(h^2).
$$

又因为

$$
\boldsymbol{D}_{u_0}^{\mathrm{T}} \boldsymbol{W}_{u_0} \begin{pmatrix} \boldsymbol{x}_1^{\mathrm{T}} \left(\dfrac{u_1 - u_0}{h} \right)^2 \\ \boldsymbol{x}_2^{\mathrm{T}} \left(\dfrac{u_2 - u_0}{h} \right)^2 \\ \vdots \\ \boldsymbol{x}_n^{\mathrm{T}} \left(\dfrac{u_n - u_0}{h} \right)^2 \end{pmatrix} = \begin{pmatrix} \displaystyle\sum_{i=1}^n \boldsymbol{x}_i \boldsymbol{x}_i^{\mathrm{T}} \left(\frac{u_i - u_0}{h} \right)^2 K_h(u_i - u_0) \\ \displaystyle\sum_{i=1}^n \boldsymbol{x}_i \boldsymbol{x}_i^{\mathrm{T}} \left(\frac{u_i - u_0}{h} \right)^3 K_h(u_i - u_0) \end{pmatrix}
$$

$$
= n\mu_2(K) f(u_0) \boldsymbol{\Gamma}(u_0) \otimes (1, 0)^{\mathrm{T}} \{1 + O_p(c_n)\}.
$$

基于上面的结论, 经过计算有

$$
\boldsymbol{SM} = \begin{pmatrix} \boldsymbol{x}_1^{\mathrm{T}} \boldsymbol{\alpha}(u_1) + \dfrac{h^2}{2} \mu_2(K) \boldsymbol{x}_1^{\mathrm{T}} \boldsymbol{\alpha}''(u_1) \\ \boldsymbol{x}_2^{\mathrm{T}} \boldsymbol{\alpha}(u_2) + \dfrac{h^2}{2} \mu_2(K) \boldsymbol{x}_2^{\mathrm{T}} \boldsymbol{\alpha}''(u_2) \\ \vdots \\ \boldsymbol{x}_n^{\mathrm{T}} \boldsymbol{\alpha}(u_n) + \dfrac{h^2}{2} \mu_2(K) \boldsymbol{x}_n^{\mathrm{T}} \boldsymbol{\alpha}''(u_n) \end{pmatrix} \{1 + O_p(c_n)\}.
$$

所以有

$$
\frac{1}{n} \boldsymbol{Z}^{\mathrm{T}} (\boldsymbol{I} - \boldsymbol{S}) \boldsymbol{M} = -\frac{h^2}{2} \mu_2(K) E(\boldsymbol{z}_i \boldsymbol{x}_i^{\mathrm{T}} \boldsymbol{\alpha}''(u_i)) + o_p(h^2). \tag{2.33}
$$

从而由式 (2.32) 与式 (2.33) 有

$$
E(\hat{\boldsymbol{\beta}}_{\mathrm{BF}} - \boldsymbol{\beta} | \boldsymbol{X}, \boldsymbol{Z}, \boldsymbol{U}) = -\frac{h^2}{2} \mu_2(K) \boldsymbol{\Xi}^{-1} E[\boldsymbol{z}_i \boldsymbol{x}_i^{\mathrm{T}} \boldsymbol{\alpha}''(u_i)] + o_p(h^2).
$$

对于 $\hat{\boldsymbol{\beta}}_{\mathrm{BF}}$ 的方差, 首先由其表达式立得

$$
\mathrm{Var}(\hat{\boldsymbol{\beta}}_{\mathrm{BF}}) = \sigma^2 [\boldsymbol{Z}^{\mathrm{T}} (\boldsymbol{I} - \boldsymbol{S}) \boldsymbol{Z}]^{-1} \boldsymbol{Z}^{\mathrm{T}} (\boldsymbol{I} - \boldsymbol{S}) (\boldsymbol{I} - \boldsymbol{S})^{\mathrm{T}} \boldsymbol{Z} [\boldsymbol{Z}^{\mathrm{T}} (\boldsymbol{I} - \boldsymbol{S})^{\mathrm{T}} \boldsymbol{Z}]^{-1}.
$$

类似于上面关于 $\boldsymbol{Z}^{\mathrm{T}} (\boldsymbol{I} - \boldsymbol{S}) \boldsymbol{Z}$ 的讨论, 同样可得到

$$
\frac{1}{n} \boldsymbol{Z}^{\mathrm{T}} (\boldsymbol{I} - \boldsymbol{S})^{\mathrm{T}} \boldsymbol{Z} \to \boldsymbol{\Xi}.
$$

再由引理 2.4 知

$$\frac{1}{n}\overline{\boldsymbol{Z}}^{\mathrm{T}}(\boldsymbol{I}-\boldsymbol{S})(\boldsymbol{I}-\boldsymbol{S})^{\mathrm{T}}\overline{\boldsymbol{Z}} \to \boldsymbol{\varXi}.$$

从而有

$$\mathrm{Var}(\hat{\boldsymbol{\beta}}_{\mathrm{BF}}|\boldsymbol{X},\boldsymbol{Z},\boldsymbol{U}) = \frac{\sigma^2}{n}\boldsymbol{\varXi}^{-1} + o_p\left(\frac{1}{n}\right).$$

证毕.

第 3 章　异方差部分线性变系数模型的研究

实际数据分析中, 观测数据的异方差性是一个非常普遍的问题, 关于此类问题的一些结果大都是基于线性回归模型进行研究所得到的. 众所周知, 对于普通线性回归模型, 当异方差存在时, 模型系数的一般最小二乘估计虽然还是相合估计, 但不再是有效估计. 考虑异方差信息的广义最小二乘估计优于一般的最小二乘估计. 对于既含有参数分量又含有非参数分量的半参数模型, 在异方差情形下如何构造模型的有效估计是重要和有意义的问题. 此外, 如何检验模型异方差的存在也是实际数据分析中要解决的问题. 本章将针对异方差部分线性变系数模型进行深入地讨论. 本章的主要内容来自于文献 Wei 和 Wu (2008) 和 Wei, Wan 和 Liu (2013).

3.1　引　　言

在经典的回归分析中, 观测值的方差齐性是一个很基本的假定, 在此假定下方可进行常规的统计推断. 如果方差非齐且未知, 则回归分析将遇到诸多问题, 从而关于异方差情形下模型的估计以及异方差的检验等问题受到了统计学家与计量经济学家的关注, 已经得到了一些非常有用的结果, 具体的内容可参考 Glejser (1969), Bickel (1978), Harrison 和 MCcabe (1979), Breusch 和 Pagan (1979), Carroll 和 Ruppert (1981), Cook 和 Weisberg (1983). 然而, 其中的大部分结果都是基于参数回归模型特别是线性回归模型的研究而得到的. 实际数据分析中, 由回归分析的经典理论可知, 回归函数的不正确设定会导致推断的严重错误, 从而基于非参数回归模型来研究异方差问题近年来受到了广泛的关注. Eubank 和 Thoms (1993) 与 Cai 等 (1998) 分别基于光滑样条方法与小波方法研究了非参数回归模型的异方差检验问题, 将在参数回归模型中得到广泛研究的得分检验方法作了推广. Dette 和 Munk (1998) 基于差分思想对一类方差函数为未知光滑函数的非参数回归模型进行了异方差检验.

虽然非参数回归方法克服了参数回归模型主观假定函数形式的严重缺陷, 充分依靠数据本身探求变量之间的关系. 但是, 非参数回归方法在处理高维数据时会遇到“维数祸根”问题. 为了解决这一问题, 近二十年来多种多元非 (半) 参数建模方法被提出用以探索变量间蕴涵的复杂关系, 比如, 可加模型、变系数模型、部分线性模型、多指标模型以及这些模型的混合形式. 近年来对于这些模型的研究大都还集中在模型的估计上, 关于异方差的检验问题还没有受到重视, 关于此类问题的结果就我们所知也非常少. 其中冉和朱 (2004) 基于光滑样条方法研究了部分线性模型的异方差检验问题, You 和 Chen (2005) 研究了一类部分线性模型的异方差检验问题, 但是其中关于方差函数的设定以及所提出的检验方法都不具有普遍性. 部分线性变系数模型作为一类广泛的半参数模型最近受到了人们的关注, 下面将对该模型在异方差情况下的估计以及异方差的检验等问题进行探讨.

变系数模型的建模思想出现在很多领域中并且已经被成功地应用到多维非参数回归模型、广义线性模型、时间序列分析、纵向数据与函数数据分析以及空间数据分析, 金融数据的时变模型中. 部分线性变系数模型作为变系数模型的推广最近受到了人们的重视, 其形式为

$$Y = \boldsymbol{X}^{\mathrm{T}}\boldsymbol{\alpha}(U) + \boldsymbol{Z}^{\mathrm{T}}\boldsymbol{\beta} + \varepsilon, \tag{3.1}$$

其中 $(U, \boldsymbol{X}^{\mathrm{T}}, \boldsymbol{Z}^{\mathrm{T}})$ 是自变量, 不失一般性, 下面的叙述中假定 U 为一维变量. Y 为因变量, $\boldsymbol{\beta} = (\beta_1, \beta_2, \cdots, \beta_q)^{\mathrm{T}}$ 为 q 维未知待估参数, $\boldsymbol{\alpha}(\cdot) = (\alpha_1(\cdot), \alpha_1(\cdot), \cdots, \alpha_p(\cdot))^{\mathrm{T}}$ 为一列未知函数. ε 为模型误差, 有 $E(\varepsilon) = 0$ 和 $\mathrm{Var}(\varepsilon) = \sigma^2(U, \boldsymbol{X}, \boldsymbol{Z})$,

接下来的 3.2 节将基于 profile 最小二乘方法构造参数分量在异方差情形下的有效估计. 3.3 节将提出针对异方差问题的 profile 得分检验统计量. 3.4 节将针对一类部分线性模型的异方差检验问题做进一步的研究.

3.2 异方差情形下参数分量的有效估计

对于异方差模型 (3.1), 如何构造参数分量 $\boldsymbol{\beta}$ 的有效估计是研究的一个重点问题. Ma 等 (2006) 针对部分线性模型提出了一类有效估计方法, 参数分量的估计量能达到半参数有效界. Ahmad 等 (2005) 针对模型 (3.1) 利用级数估计方法研究了异方差下有效估计量的构造问题, 这篇文章也指出基于核方法 (局部多项式) 为基础构造参数分量的有效估计是有难度的. 为了解决这一问题, 下面将基于 Fan 和 Huang (2005) 提出的基于局部线性方法的 profile 最小二乘技术来构造参数分量的有效估计. 为了叙述的连贯性, 首先简要回顾模型 (3.1) 在不考虑异方差情形下的 Proflie 最小二乘估计.

假设有对模型 (3.1) 的 n 组观测值为 $\{(u_k, x_{k1}, \cdots, x_{kp}, z_{k1}, \cdots, z_{kq}, y_k)\}, k = 1, 2, \cdots, n$. 引入与第 2 章相同意义的符号, 即

$$\boldsymbol{X} = \begin{pmatrix} \boldsymbol{x}_1^{\mathrm{T}} \\ \boldsymbol{x}_2^{\mathrm{T}} \\ \vdots \\ \boldsymbol{x}_n^{\mathrm{T}} \end{pmatrix} = \begin{pmatrix} x_{11} & \cdots & x_{1p} \\ x_{21} & \cdots & x_{2p} \\ \vdots & & \vdots \\ x_{n1} & \cdots & x_{np} \end{pmatrix}, \quad \boldsymbol{Z} = \begin{pmatrix} \boldsymbol{z}_1^{\mathrm{T}} \\ \boldsymbol{z}_2^{\mathrm{T}} \\ \vdots \\ \boldsymbol{z}_n^{\mathrm{T}} \end{pmatrix} = \begin{pmatrix} z_{11} & \cdots & z_{1q} \\ z_{21} & \cdots & z_{2q} \\ \vdots & & \vdots \\ z_{n1} & \cdots & z_{nq} \end{pmatrix},$$

$$\boldsymbol{M}=\begin{pmatrix}\alpha_1(u_1)x_{11}+\cdots+\alpha_p(u_1)x_{1p}\\ \alpha_1(u_2)x_{21}+\cdots+\alpha_p(u_2)x_{2p}\\ \vdots\\ \alpha_1(u_n)x_{n1}+\cdots+\alpha_p(u_n)x_{np}\end{pmatrix}=\begin{pmatrix}\boldsymbol{\alpha}^{\mathrm{T}}(u_1)\boldsymbol{x}_1\\ \boldsymbol{\alpha}^{\mathrm{T}}(u_2)\boldsymbol{x}_2\\ \vdots\\ \boldsymbol{\alpha}^{\mathrm{T}}(u_n)\boldsymbol{x}_n\end{pmatrix},$$

$$\boldsymbol{D}_{u_0}=\begin{pmatrix}\boldsymbol{x}_1^{\mathrm{T}} & \boldsymbol{x}_1^{\mathrm{T}}\dfrac{u_1-u_0}{h}\\ \boldsymbol{x}_2^{\mathrm{T}} & \boldsymbol{x}_2^{\mathrm{T}}\dfrac{u_2-u_0}{h}\\ \vdots & \vdots\\ \boldsymbol{x}_n^{\mathrm{T}} & \boldsymbol{x}_n^{\mathrm{T}}\dfrac{u_n-u_0}{h}\end{pmatrix},$$

以及 $\boldsymbol{Y}=(y_1,y_2,\cdots,y_n)^{\mathrm{T}}$, $\boldsymbol{W}_{u_0}=\mathrm{diag}(K_h(u_1-u_0),K_h(u_1-u_0),\cdots,K_h(u_n-u_0))$, 局部线性光滑矩阵

$$\boldsymbol{S}=\begin{pmatrix}(\boldsymbol{x}_1^{\mathrm{T}}\ \boldsymbol{0})\{\boldsymbol{D}_{u_1}^{\mathrm{T}}\boldsymbol{W}_{u_1}\boldsymbol{D}_{u_1}\}^{-1}\boldsymbol{D}_{u_1}^{\mathrm{T}}\boldsymbol{W}_{u_1}\\ (\boldsymbol{x}_2^{\mathrm{T}}\ \boldsymbol{0})\{\boldsymbol{D}_{u_2}^{\mathrm{T}}\boldsymbol{W}_{u_2}\boldsymbol{D}_{u_2}\}^{-1}\boldsymbol{D}_{u_2}^{\mathrm{T}}\boldsymbol{W}_{u_2}\\ \vdots\\ (\boldsymbol{x}_n^{\mathrm{T}}\ \boldsymbol{0})\{\boldsymbol{D}_{u_n}^{\mathrm{T}}\boldsymbol{W}_{u_n}\boldsymbol{D}_{u_n}\}^{-1}\boldsymbol{D}_{u_n}^{\mathrm{T}}\boldsymbol{W}_{u_n}\end{pmatrix}.$$

由 2.2 节, 参数分量 $\boldsymbol{\beta}$ 的 profile 最小二乘估计为

$$\hat{\boldsymbol{\beta}}_{\mathrm{PL}}=[\boldsymbol{Z}^{\mathrm{T}}(\boldsymbol{I}-\boldsymbol{S})^{\mathrm{T}}(\boldsymbol{I}-\boldsymbol{S})\boldsymbol{Z}]^{-1}\boldsymbol{Z}^{\mathrm{T}}(\boldsymbol{I}-\boldsymbol{S})^{\mathrm{T}}(\boldsymbol{I}-\boldsymbol{S})\boldsymbol{Y}, \tag{3.2}$$

相应地, 变系数部分 $\boldsymbol{M}$ 的估计为

$$\hat{\boldsymbol{M}}=\boldsymbol{S}(\boldsymbol{Y}-\boldsymbol{Z}\hat{\boldsymbol{\beta}}_{\mathrm{PL}}), \tag{3.3}$$

因变量 $\boldsymbol{Y}$ 的拟合值为

$$\hat{\boldsymbol{Y}}=\hat{\boldsymbol{M}}+\boldsymbol{Z}\hat{\boldsymbol{\beta}}=\boldsymbol{L}\boldsymbol{Y}, \tag{3.4}$$

基于此拟合方法的残差向量为

$$\hat{\boldsymbol{\varepsilon}}=(\hat{\varepsilon}_1,\hat{\varepsilon}_2,\cdots,\hat{\varepsilon}_n)^{\mathrm{T}}=\boldsymbol{Y}-\hat{\boldsymbol{Y}}=(\boldsymbol{I}-\boldsymbol{L})\boldsymbol{Y}. \tag{3.5}$$

其中 $\boldsymbol{L}=\mathbf{S}+(\boldsymbol{I}-\boldsymbol{S})\boldsymbol{Z}[\boldsymbol{Z}^{\mathrm{T}}(\boldsymbol{I}-\boldsymbol{S})^{\mathrm{T}}(\boldsymbol{I}-\boldsymbol{S})\boldsymbol{Z}]^{-1}\boldsymbol{Z}^{\mathrm{T}}(\boldsymbol{I}-\boldsymbol{S})^{\mathrm{T}}(\boldsymbol{I}-\boldsymbol{S})$.

对于 profile 最小二乘估计 $\hat{\boldsymbol{\beta}}_{\mathrm{PL}}$, Fan 和 Huang (2005) 在方差齐性 (即 $\sigma^2(\boldsymbol{x}_i,\boldsymbol{z}_i,u_i)=\sigma^2, i=1,2,\cdots,n$) 的条件下给出了其渐近性质. 在异方差情形下有下面的结果. 下面定理成立的一些条件将在 3.5 节中给出, 记 $\Gamma(U)=E(\boldsymbol{X}\boldsymbol{X}^{\mathrm{T}}|U)$, $\varPhi(U)=E(\boldsymbol{X}\boldsymbol{Z}^{\mathrm{T}}|U)$.

定理 3.1 若满足 3.5 节中的条件 (A.1)~ 条件 (A.6), 那么基于异方差模型 (3.1) 得到的 $\boldsymbol{\beta}$ 的 profile 最小二乘估计是渐近正态的, 即

$$\sqrt{n}(\hat{\boldsymbol{\beta}}_{\mathrm{PL}}-\boldsymbol{\beta})\to N(\mathbf{0},\boldsymbol{\Sigma}),$$

其中 $\boldsymbol{\Sigma}=\boldsymbol{\Xi}^{-1}\boldsymbol{\Omega}\boldsymbol{\Xi}^{-1}$, $\boldsymbol{\Xi}=E[e_ie_i^{\mathrm{T}}]$, $\boldsymbol{\Omega}=E[e_ie_i^{\mathrm{T}}\sigma^2(\boldsymbol{x}_i,\boldsymbol{z}_i,u_i)]$ 和 $\sigma^2(\boldsymbol{x}_i,\boldsymbol{z}_i,u_i)=E[\varepsilon_i^2|\boldsymbol{x}_i,\boldsymbol{z}_i,u_i], e_i=\boldsymbol{z}_i-\Phi^{\mathrm{T}}(u_i)\Gamma^{-1}(u_i)\boldsymbol{x}_i$.

显然, 如果模型 (3.1) 满足方差齐性, 即有 $\sigma^2(\boldsymbol{x}_i,\boldsymbol{z}_i,u_i)=\sigma^2, i=1,2,\cdots,n$, 则有

$$\boldsymbol{\Omega}=E[e_ie_i^{\mathrm{T}}\sigma^2]=\sigma^2\boldsymbol{\Xi},\quad \boldsymbol{\Sigma}=\boldsymbol{\Xi}^{-1}\boldsymbol{\Omega}\boldsymbol{\Xi}^{-1}=\sigma^2\boldsymbol{\Xi}^{-1},$$

从而有

$$\sqrt{n}(\hat{\boldsymbol{\beta}}-\boldsymbol{\beta})\to N(\mathbf{0},\boldsymbol{\sigma^2\Xi^{-1}}).$$

该结论即为 Fan 和 Huang (2005) 中的定理 4.1. 所以上面的结论可以看成 Fan 和 Huang (2005) 中的定理 4.1 的一个推广.

基于上面的定理, 可以进行相关的统计推断, 但是 $\hat{\beta}_{\mathrm{PL}}$ 的渐近协方差矩阵 $\boldsymbol{\Sigma}$ 是未知的, 类似于线性回归模型, 构造如下的对异方差情形稳健的估计量

$$\hat{\boldsymbol{\Sigma}}=\left(\frac{1}{n}\sum_{i=1}^n\widetilde{\boldsymbol{z}}_i\widetilde{\boldsymbol{z}}_i^{\mathrm{T}}\right)^{-1}\left(\frac{1}{n}\sum_{i=1}^n\widetilde{\boldsymbol{z}}_i\widetilde{\boldsymbol{z}}_i^{\mathrm{T}}\hat{\varepsilon}_i^2\right)\left(\frac{1}{n}\sum_{i=1}^n\widetilde{\boldsymbol{z}}_i\widetilde{\boldsymbol{z}}_i^{\mathrm{T}}\right)^{-1},$$

其中 $\widetilde{\boldsymbol{Z}}=(\widetilde{z}_1,\widetilde{z}_2,\cdots,\widetilde{z}_n)^{\mathrm{T}}=(\boldsymbol{I}-\boldsymbol{S})\boldsymbol{Z}$.

定理 3.2 若 2.5 节中的条件 (A.1)~ 条件 (A.6) 成立, 则有

$$\hat{\boldsymbol{\Sigma}}\xrightarrow{P}\boldsymbol{\Sigma},$$

从而对于 $\hat{\boldsymbol{\beta}}_{\mathrm{PL}}$ 有

$$\sqrt{n}\hat{\boldsymbol{\Sigma}}^{-1/2}(\hat{\boldsymbol{\beta}}_{\mathrm{PL}}-\boldsymbol{\beta})\to N(\mathbf{0},\boldsymbol{I}_{\mathbf{q}}).$$

该结论可用于关于 $\boldsymbol{\beta}$ 的大样本置信椭球估计以及假设检验. 比如可以构造 $r^{\mathrm{T}}\boldsymbol{\beta}$ 的置信度为 0.95 的置信区间 $r^{\mathrm{T}}\hat{\boldsymbol{\beta}}_{\mathrm{PL}}\pm 1.96\sqrt{r^{\mathrm{T}}\hat{\boldsymbol{\Sigma}}^{-1}r}$. 对于线性检验问题 $H_0:\boldsymbol{A}\boldsymbol{\beta}=\boldsymbol{b}$, 可以构造如下的 Wald 检验统计量

$$\mathrm{T}=(\boldsymbol{A}\hat{\boldsymbol{\beta}}_{\mathrm{PL}}-\boldsymbol{b})^{\mathrm{T}}\left(\boldsymbol{A}\hat{\boldsymbol{\Sigma}}\boldsymbol{A}^{\mathrm{T}}\right)^{-1}(\boldsymbol{A}\hat{\boldsymbol{\beta}}_{\mathrm{PL}}-\boldsymbol{b}).$$

该统计量在原假设成立的情况下其渐近分布为自由度为 k 的 χ^2 分布.

前面给出的 profile 最小二乘估计没有考虑异方差, 为了得到更有效的估计, 我们尝试着构造几类不同的估计. 首先我们构造 $\boldsymbol{\beta}$ 的加权估计.

记 $\sigma_i = \sqrt{\sigma^2(\boldsymbol{x}_i, \boldsymbol{z}_i, u_i)}$, 由第 2 章知道, $\boldsymbol{\beta}$ 的 profile 最小二乘估计是基于下面的线性模型 (见式 (2.9)) 得到的.

$$\widetilde{y}_i = \boldsymbol{\beta}^{\mathrm{T}} \widetilde{\boldsymbol{z}}_i + \varepsilon_i,$$

其中 $\widetilde{\boldsymbol{Y}} = (\widetilde{y}_1, \widetilde{y}_2, \cdots, \widetilde{y}_n)^{\mathrm{T}} = (\boldsymbol{I} - \boldsymbol{S})\boldsymbol{Y}$. 可将上面的线性模型做如下转换

$$\widetilde{y}_i / \sigma_i = \boldsymbol{\beta}^{\mathrm{T}} \widetilde{\boldsymbol{z}}_i / \sigma_i + \varepsilon_i / \sigma_i, \tag{3.6}$$

则该模型满足方差齐性, 从而基于该模型可得 $\boldsymbol{\beta}$ 的加权 profile 最小二乘估计为

$$\hat{\boldsymbol{\beta}}_{\mathrm{GPL}} = \left(\sum_{i=1}^{n} \widetilde{\boldsymbol{z}}_i \widetilde{\boldsymbol{z}}_i^{\mathrm{T}} / \sigma_i^2 \right)^{-1} \left(\sum_{i=1}^{n} \widetilde{\boldsymbol{z}}_i \widetilde{y}_i / \sigma_i^2 \right). \tag{3.7}$$

该估计具有如下的性质.

定理 3.3 若 3.5 节中的条件 (A.1)~ 条件 (A.6) 成立, $\boldsymbol{\beta}$ 的加权 Profiel 最小二乘估计是渐近正态的, 即

$$\sqrt{n}(\hat{\boldsymbol{\beta}}_{\mathrm{GPL}} - \boldsymbol{\beta}) \to N(\boldsymbol{0}, \boldsymbol{\Phi}^{-1}),$$

其中 $\boldsymbol{\Phi} = E[e_i e_i^{\mathrm{T}} / \sigma^2(\boldsymbol{x}_i, \boldsymbol{z}_i, u_i)]$.

然而, $\hat{\boldsymbol{\beta}}_{\mathrm{GPL}}$ 并不能达到半参数有效界, 下面构造另外一种能达到半参数有效界的估计. 首先假定 σ_i 已知. 基于观测值, 模型 (3.1) 可记为

$$y_i = \boldsymbol{x}_i^{\mathrm{T}} \boldsymbol{\alpha}(u_i) + \boldsymbol{z}_i^{\mathrm{T}} \boldsymbol{\beta} + \varepsilon_i, \quad i = 1, 2, \cdots, n.$$

对于上面的模型两边同乘以 $1/\sigma_i$, 可以得到如下的模型

$$y_i^* = \boldsymbol{x}_i^{*\mathrm{T}} \boldsymbol{\alpha}(u_i) + \boldsymbol{z}_i^{*\mathrm{T}} \boldsymbol{\beta} + e_i, \quad i = 1, 2, \cdots, n,$$

其中 $y_i^* = y_i/\sigma_i, \boldsymbol{x}_i^* = \boldsymbol{x}_i/\sigma_i, \boldsymbol{z}_i^* = \boldsymbol{z}_i/\sigma_i$, 新的模型误差 $e_i = \varepsilon_i/\sigma_i$ 有 $E(e_i|\boldsymbol{x}_i, \boldsymbol{z}_i, u_i) = 0, \mathrm{Var}(e_i|\boldsymbol{x}_i, \boldsymbol{z}_i, u_i) = 1$. 显然, 转换后的模型为同方差的部分线性变系数模型. 对于这个变换后的同方差部分线性变系数模型, 利用 profile 最小二乘法, 可得到 $\boldsymbol{\beta}$ 的估计, 记为 $\widetilde{\boldsymbol{\beta}}$. 直接利用定理 3.1, 便可得到该估计量的性质如下.

定理 3.4 若 3.5 节中的条件 (A.1)~ 条件 (A.6) 成立, $\boldsymbol{\beta}$ 的有效估计 $\widetilde{\boldsymbol{\beta}}$ 是渐近正态的,

$$\sqrt{n}(\widetilde{\boldsymbol{\beta}} - \boldsymbol{\beta}) \to N(\boldsymbol{0}, \boldsymbol{\Delta}^{-1}),$$

其中$\boldsymbol{\Delta} = E\left[\frac{\boldsymbol{z}_i \boldsymbol{z}_i^{\mathrm{T}}}{\sigma^2(\boldsymbol{x}_i, \boldsymbol{z}_i, u_i)}\right] - E\Bigg\{E\left[\frac{\boldsymbol{z}_i \boldsymbol{x}_i^{\mathrm{T}}}{\sigma^2(\boldsymbol{x}_i, \boldsymbol{z}_i, u_i)}|u_i\right] E\left[\frac{\boldsymbol{x}_i \boldsymbol{x}_i^{\mathrm{T}}}{\sigma^2(\boldsymbol{x}_i, \boldsymbol{z}_i, u_i)}|u_i\right]^{-1}$

$$E\left[\frac{\boldsymbol{x}_i\boldsymbol{z}_i^{\mathrm{T}}}{\sigma^2(\boldsymbol{x}_i,\boldsymbol{z}_i,u_i)}|u_i\right]\Bigg\}.$$

注 3.1 Chamberlain (1992) 对多类半参数模型定义了半参数有效界. 由 Chamberlain (1992) 中的结果, 易知 $\boldsymbol{\Delta}$ 是部分线性变系数模型中参数分量 $\boldsymbol{\beta}$ 的有效界. 因此, 估计量 $\widetilde{\boldsymbol{\beta}}$ 是半参数有效估计.

注 3.2 如果 $p=1, \boldsymbol{X}=1$, 部分线性变系数模型转变化部分线性模型, 同时有

$$\boldsymbol{\Delta}=\mathbb{E}\left\{\frac{\boldsymbol{z}_i\boldsymbol{z}_i^{\mathrm{T}}}{\sigma^2(\boldsymbol{z}_i,u_i)}-E\left[\frac{\boldsymbol{z}_i}{\sigma^2(\boldsymbol{z}_i,u_i)}|u_i\right]E\left[\frac{1}{\sigma^2(\boldsymbol{z}_i,u_i)}|u_i\right]^{-1}E\left[\frac{\boldsymbol{z}_i^{\mathrm{T}}}{\sigma^2(\boldsymbol{z}_i,u_i)}|u_i\right]\right\}.$$

$\boldsymbol{\Delta}$ 就是 Chamberlain(1992)[569] 式 (1.9) 和 Ma 等 (2006)[78] 页式 (8) 针对部分线性模型得到了半参数有效界. 同 Chamberlain (1992) 和 Ma 等 (2006) 构造有效估计的方法相比, 我们所用的方法相对来说是直接和简单的.

注 3.3 我们所构造的 $\widetilde{\boldsymbol{\beta}}$ 依赖于 $\sigma^2(\boldsymbol{x},\boldsymbol{z},u)$. 而 $\sigma^2(\boldsymbol{x},\boldsymbol{z},u)$ 在实际问题分析中往往是未知的. 为了解决这一问题, 需要去估计未知的 $\sigma^2(\boldsymbol{x},\boldsymbol{z},u)$, 然后利用其相合估计值代替 $\widetilde{\boldsymbol{\beta}}$ 中未知的 $\sigma^2(\boldsymbol{x},\boldsymbol{z},u)$. 很容易证明由此得到的估计量的渐近性质与 $\widetilde{\boldsymbol{\beta}}$ 的一样.

注 3.4 为了得到 $\sigma^2(\boldsymbol{x},\boldsymbol{z},u)$ 的相合估计, 如果将其看成未知待估参数, 则模型估计变得非常复杂, 甚至可能是“不可识别”的. 实际应用中, 目前较为普遍的做法是将其参数化或者非参数化. 关于参数形式的方差函数的讨论可参考 Carroll 和 Ruppert (1988). Muller 和 Stadtmuller (1987), Chiou 和 Muller (1999), Ruppert 等 (1997) 讨论了非参数形式的方差函数. Muller 和 Zhao (1995) 和 Keilegom 和 Wang (2010) 讨论了半参数形式的方差函数.

数值模拟 下面通过数值模拟来考察前面针对参数分量所提出的有效估计方法. 考虑如下的异方差部分线性变系数模型

$$y_i=x_i\alpha(u_i)+z_i\beta+\sigma(x_i,z_i,u_i)\varepsilon_i,\quad i=1,2,\cdots,n, \tag{3.8}$$

其中 $x_i\sim N(1,1), u_i=i/n, z_i\sim U(-1,1), \alpha(u_i)=u_i+\sin(2\pi u_i), \beta=1$. 首先考察如下四种类型的方差函数.

$$\begin{aligned}&(\mathrm{A}):\sigma(x_i,z_i,u_i)=e^{x_i}; &&(\mathrm{B}):\sigma(x_i,z_i,u_i)=e^{u_i};\\&(\mathrm{C}):\sigma(x_i,z_i,u_i)=1+x_i; &&(\mathrm{D}):\sigma(x_i,z_i,u_i)=1+z_i.\end{aligned}$$

此外, 为了考察误差的分布对结果的影响, 考察如下四种误差分布①$\varepsilon_i\sim N(0,0.5^2)$, ②$\varepsilon_i\sim U(-\sqrt{3}/2,\sqrt{3}/2)$, ③$\varepsilon_i\sim\dfrac{1}{8}\chi_8^2-1$. 模拟中采用高斯核函数, 为了计算方便, 采取 $h=n^{-1/5}$, 样本量 $n=30,50,80$. 基于上面设定的各种情况, 重复模拟 1000

次, 每次利用前面所提到的估计方法可得到不考虑异方差的一般 profile 最小二乘估计 $\hat{\beta}$ 和有效估计 $\widetilde{\beta}$. 针对每种估计值, 计算基于 1000 次估计值得到的偏差 (Bias) 和标准差 (SD) 以及经验均方误差 (MSE), 具体结果见表 3.1~ 表 3.4. 不难看出在各种情形下, 有效估计 $\widetilde{\beta}$ 都优于一般估计.

表 3.1　A 类方差函数下 $\hat{\beta}$ 与 $\widetilde{\beta}$ 的比较

误差	n	$\hat{\beta}$			$\widetilde{\beta}$		
		Bias	SD	MSE	Bias	SD	MSE
$N(0,0.5^2)$	30	0.032	0.905	0.819	0.005	0.290	0.084
	50	0.012	0.850	0.721	−0.008	0.209	0.043
	80	−0.005	0.613	0.375	0.005	0.150	0.022
$U\left(-\frac{\sqrt{3}}{2},\frac{\sqrt{3}}{2}\right)$	30	−0.080	0.981	0.967	0.001	0.278	0.077
	50	0.056	0.804	0.648	0.015	0.203	0.041
	80	−0.026	0.645	0.416	−0.003	0.148	0.021
$\frac{1}{8}\chi_8^2-1$	30	−0.038	0.833	0.694	−0.004	0.309	0.095
	50	0.013	0.822	0.675	0.012	0.206	0.042
	80	−0.034	0.628	0.395	−0.002	0.149	0.022

表 3.2　B 类方差函数下 $\hat{\beta}$ 与 $\widetilde{\beta}$ 的比较

误差	n	$\hat{\beta}$			$\widetilde{\beta}$		
		Bias	SD	MSE	Bias	SD	MSE
$N(0,0.5^2)$	30	0.010	0.329	0.108	0.009	0.284	0.081
	50	−0.005	0.246	0.060	−0.002	0.219	0.048
	80	0.002	0.170	0.029	0.002	0.150	0.022
$U\left(-\frac{\sqrt{3}}{2},\frac{\sqrt{3}}{2}\right)$	30	0.024	0.331	0.110	0.012	0.288	0.083
	50	−0.009	0.239	0.057	−0.019	0.204	0.042
	80	0.002	0.182	0.033	0.003	0.157	0.024
$\frac{1}{8}\chi_8^2-1$	30	0.011	0.312	0.097	0.013	0.279	0.077
	50	0.017	0.226	0.051	0.016	0.205	0.042
	80	−0.008	0.186	0.034	−0.012	0.156	0.024

表 3.3　C 类方差函数下 $\hat{\beta}$ 与 $\widetilde{\beta}$ 的比较

误差	n	$\hat{\beta}$			$\widetilde{\beta}$		
		Bias	SD	MSE	Bias	SD	MSE
$N(0,0.5^2)$	30	0.024	0.348	0.121	0.022	0.239	0.057
	50	0.001	0.282	0.079	0.007	0.189	0.036
	80	0.009	0.215	0.046	−0.001	0.130	0.017

续表

误差	n	$\hat{\beta}$			$\widetilde{\beta}$		
		Bias	SD	MSE	Bias	SD	MSE
$U\left(-\frac{\sqrt{3}}{2},\frac{\sqrt{3}}{2}\right)$	30	−0.003	0.368	0.135	0.002	0.251	0.063
	50	−0.005	0.281	0.078	−0.003	0.171	0.029
	80	0.005	0.212	0.045	0.000	0.136	0.018
$\frac{1}{8}\chi_8^2-1$	30	−0.013	0.380	0.144	−0.002	0.257	0.066
	50	0.001	0.279	0.078	0.015	0.170	0.029
	80	−0.008	0.222	0.049	0.000	0.132	0.017

表 3.4　D 类方差函数下 $\hat{\beta}$ 与 $\widetilde{\beta}$ 的比较

误差	n	$\hat{\beta}$			$\widetilde{\beta}$		
		Bias	SD	MSE	Bias	SD	MSE
$N(0,0.5^2)$	30	0.009	0.208	0.043	−0.002	0.167	0.028
	50	−0.011	0.164	0.027	0.002	0.107	0.011
	80	0.003	0.126	0.015	0.005	0.095	0.009
$U\left(-\frac{\sqrt{3}}{2},\frac{\sqrt{3}}{2}\right)$	30	−0.004	0.220	0.048	0.002	0.191	0.036
	50	−0.006	0.168	0.028	0.001	0.107	0.011
	80	0.000	0.125	0.015	0.006	0.078	0.006
$\frac{1}{8}\chi_8^2-1$	30	−0.011	0.212	0.045	−0.005	0.174	0.030
	50	0.014	0.155	0.024	0.012	0.123	0.015
	80	0.008	0.124	0.015	−0.001	0.086	0.007

前面的模拟采用的是已知的方差函数, 下面考察方差函数未知的情况. 为了方便, 考虑如下的方差函数

$$E(\varepsilon_i^2|\boldsymbol{x}_i,\boldsymbol{z}_i,u_i)=\sigma^2(\boldsymbol{x}_i,\boldsymbol{z}_i,u_i)=\exp(\gamma_0+\gamma_1u_i).$$

显然, 基于上式, 我们可以构造如下的回归模型

$$\ln\varepsilon_i^2=\gamma_0+\gamma_1u_i+\xi_i,$$

其中 $E\xi_i=0$. 由于 ε_i 是未知的, 我们便用其估计值 $\hat{\varepsilon}_i=y_i-\boldsymbol{x}_i^{\mathrm{T}}\hat{\boldsymbol{\alpha}}(u_i)-\boldsymbol{z}_i^{\mathrm{T}}\hat{\boldsymbol{\beta}}$ 来代替, 其中 $\hat{\boldsymbol{\beta}}$ 和 $\hat{\boldsymbol{\alpha}}(u_i)$ 是不考虑异方差下的一般估计. 将 ε_i 用 $\hat{\varepsilon}_i$ 代替, 并利用最小二乘法来估计上述线性模型, 可得到 γ_0,γ_1 的估计, 记为 $\hat{\gamma}_0$ 和 $\hat{\gamma}_1$. 从而我们可得到 $\sigma^2(\boldsymbol{x}_i,\boldsymbol{z}_i,u_i)$ 的估计 $\hat{\sigma}^2(\boldsymbol{x}_i,\boldsymbol{z}_i,u_i)=\exp(\hat{\gamma}_0+\hat{\gamma}_1u_i)$. 与前面的模拟一样, 我们可得到模拟结果表 3.5, 不难看出在此情形下有效估计 $\widetilde{\beta}$ 也优于一般估计 $\hat{\beta}$.

表 3.5　方差函数未知时 $\hat{\beta}$ 与 $\widetilde{\beta}$ 的比较

误差	n	$\hat{\beta}$			$\widetilde{\beta}$		
		Bias	SD	MSE	Bias	SD	MSE
$N(0,0.5^2)$	30	−0.099	1.376	1.902	−0.069	1.272	1.623
	50	0.058	1.059	1.124	0.027	0.945	0.893
	80	0.019	0.795	0.632	0.014	0.711	0.505
$U\left(-\frac{\sqrt{3}}{2},\frac{\sqrt{3}}{2}\right)$	30	0.045	1.392	1.937	0.028	1.294	1.674
	50	−0.016	1.018	1.036	−0.021	0.894	0.799
	80	−0.002	0.794	0.630	−0.004	0.701	0.491
$\frac{1}{8}\chi_8^2-1$	30	0.015	1.356	1.837	0.009	1.254	1.573
	50	0.026	1.061	1.127	0.028	0.958	0.918
	80	−0.012	0.820	0.672	−0.014	0.722	0.521

3.3　profile 得分检验统计量的构造及性质

如果模型 (3.1) 满足方差齐性, 则可对其进行常规的统计推断, 若是方差非齐, 则要考虑进行数据变换以及其他处理方法, 比如上面提到的方法. 因此, 检验数据的异方差性是否存在是处理问题的第一步, 它在理论与应用上都是十分重要的问题. 文献中关于异方差检验的结果, 大都是基于参数回归模型特别是线性回归模型研究所得到的. 下面基于 Cook 和 Weisbeg (1983) 的设定方法研究部分线性变系数模型的异方差检验问题.

考虑如下的异方差部分线性变系数模型

$$y_i=\boldsymbol{\alpha}^{\mathrm{T}}(u_i)\boldsymbol{x}_i+\boldsymbol{\beta}^{\mathrm{T}}\boldsymbol{z}_i+\sqrt{\sigma^2W_i}\varepsilon_i,\quad i=1,2,\cdots,n, \tag{3.9}$$

其中 $\varepsilon_i(i=1,2,\cdots,n)$ 独立同分布, 有 $E(\varepsilon_1)=0$ 和 $\mathrm{Var}(\varepsilon_1)=1$. 其余符号同式 (3.1). 依据 Cook 和 Weisbeg (1983), 假定 W_i 与某一协变量 R_i 和某一 k 维未知参数 θ 有关,

$$W_i=W(R_i,\boldsymbol{\theta}), \tag{3.10}$$

其中 $W(\cdot,\cdot)$ 为已知函数, 常被称为权函数, 其中的 $\boldsymbol{\theta}$ 为结构参数, 且存在唯一的 $\boldsymbol{\theta}_0$, 使得对一切 i 都有 $W(R_i,\theta_0)=1$. Cook 和 Weisbeg (1983) 还给出了利用图形方法构造方差权函数的可能形式. 若模型的残差 (或学生化残差) 对因边量 y_i 或者因变量 y_i 的拟合值的散点图呈喇叭状, 则 $W(\cdot,\cdot)$ 可取为 $f(x_i,\boldsymbol{\beta})$ 的函数的形式; 若模型的残差 (或学生化残差) 对某协变量 R_i 的散点图呈喇叭状, 则 $W(\cdot,\cdot)$ 可取为 R_i 与 $\boldsymbol{\theta}$ 的函数形式, 包括对数线性模型 $W(R_i,\boldsymbol{\theta})=\exp\{R_i^{\mathrm{T}}\boldsymbol{\theta}\}$ 与幂积模型

$W(R_i,\boldsymbol{\theta}) = \prod_{j=1}^{l} R_{ij}^{\boldsymbol{\theta}_j}$, 这两种模型是方差参数结构化的常用形式. 这时候模型 (3.9) 异方差的检验可化为一般的参数假设检验问题

$$H_0: \quad \theta = \theta_0 \qquad \text{VS} \qquad H_1: \quad \theta \neq \theta_0, \tag{3.11}$$

显然, 如果原假设成立, 那么模型具有方差齐性, 否则则认为存在异方差现象.

基于线性回归模型 (即模型 (3.9) 中 $\boldsymbol{x}_i = \boldsymbol{0}$), Cook 和 Weisbeg(1983) 在上面的假定下构造了得分检验统计量, 该检验方法受到了人们的重视, 并被推广到许多模型. 得分统计量的最大优点是, 构造过程中只需要计算原假设成立时 (即方差齐性) 参数的极大似然估计, 而不需要计算备择假设下 (即异方差条件下) 参数的极大似然估计, 而且其渐近分布与似然比统计量相同, 检验的功效也相当. 然而对于半参数模型 (3.9), 模型中含有未知函数 $\boldsymbol{\alpha}(\cdot)$, 模型的极大似然估计不存在或者求解非常复杂. 为了使得提出的方法具有广泛的适用性, 下面基于 profile 最小二乘估计来研究异方差的检验问题.

假定 $\boldsymbol{\varepsilon} \sim N(\boldsymbol{0}, \sigma^2 \boldsymbol{I}_n)$, 那么模型 (3.9) 的对数似然函数可写为

$$L(\boldsymbol{\alpha},\boldsymbol{\beta},\sigma^2,\theta) = c - \frac{n}{2}\log\sigma^2 - \frac{1}{2}\sum_{i=1}^{n}\log W(R_i,\theta) - \sum_{i=1}^{n}\frac{(y_i - \boldsymbol{\alpha}^{\mathrm{T}}(u_i)\boldsymbol{x}_i - \boldsymbol{\beta}^{\mathrm{T}}\boldsymbol{z}_i)^2}{2\sigma^2 W(R_i,\theta)}, \tag{3.12}$$

其中 c 为一常数. 给定 $\boldsymbol{\beta}$, 那么 $\boldsymbol{\alpha}(\cdot)$ 可以利用局部线性方法得到其估计为 $\hat{\boldsymbol{\alpha}}(\cdot;\boldsymbol{\beta})$, 将该估计代入上面的似然函数, 则得到如下的 profile 似然函数

$$\begin{aligned} L(\hat{\boldsymbol{\alpha}}(\cdot;\boldsymbol{\beta}),\boldsymbol{\beta},\sigma^2,\boldsymbol{\theta}) &= c - \frac{n}{2}\log\sigma^2 - \frac{1}{2}\sum_{i=1}^{n}\log W(R_i,\theta) - \sum_{i=1}^{n}\frac{(y_i - \hat{\boldsymbol{\alpha}}^{\mathrm{T}}(u_i)\boldsymbol{x}_i - \boldsymbol{\beta}^{\mathrm{T}}\boldsymbol{z}_i)^2}{2\sigma^2 W(R_i,\theta)} \\ &= c - \frac{n}{2}\log\sigma^2 - \frac{1}{2}\sum_{i=1}^{n}\log W(R_i,\theta) - \sum_{i=1}^{n}\frac{(\widetilde{y}_i - \boldsymbol{\beta}^{\mathrm{T}}\widetilde{\boldsymbol{z}}_i)^2}{2\sigma^2 W(R_i,\theta)}. \end{aligned} \tag{3.13}$$

如果原假设成立, 即模型满足方差齐性, 上面的 profile 似然函数转化为

$$L(\hat{\boldsymbol{\alpha}}(\cdot;\boldsymbol{\beta}),\boldsymbol{\beta},\sigma^2) = c - \frac{n}{2}\log\sigma^2 - \sum_{i=1}^{n}\frac{(\widetilde{y}_i - \widetilde{\boldsymbol{z}}_i^{\mathrm{T}}\boldsymbol{\beta})^2}{2\sigma^2}, \tag{3.14}$$

则由 $\dfrac{\partial L(\boldsymbol{\alpha},\boldsymbol{\beta},\sigma^2)}{\partial \boldsymbol{\beta}} = 0$ 可得 $\boldsymbol{\beta}$ 的 proflile 似然估计为

$$\hat{\boldsymbol{\beta}}_{\mathrm{PM}} = [\boldsymbol{Z}^{\mathrm{T}}(\boldsymbol{I}-\boldsymbol{S})^{\mathrm{T}}(\boldsymbol{I}-\boldsymbol{S})\boldsymbol{Z}]^{-1}\boldsymbol{Z}^{\mathrm{T}}(\boldsymbol{I}-\boldsymbol{S})^{\mathrm{T}}(\boldsymbol{I}-\boldsymbol{S})\boldsymbol{Y}. \tag{3.15}$$

显然, 该估计与 profile 最小二乘估计相同. 由 $\dfrac{\partial L(\boldsymbol{\alpha},\boldsymbol{\beta},\sigma^2,)}{\partial \sigma^2} = 0$ 可得 σ^2 的 proflile 似然估计为

$$\hat{\sigma}^2_{\mathrm{PM}} = n^{-1}\sum_{i=1}^{n}(\widetilde{y}_i - \hat{\boldsymbol{\beta}}_{\mathrm{PM}}^{\mathrm{T}}\widetilde{\boldsymbol{z}}_i)^2. \tag{3.16}$$

值得注意的是由于 $\hat{\boldsymbol{\alpha}}(\cdot;\boldsymbol{\beta})$ 不是极大似然方法估计所得, 所以 $\hat{\boldsymbol{\beta}}_{\mathrm{PM}}$ 与 $\hat{\sigma}^2_{\mathrm{PM}}$ 不是极大似然估计.

下面基于 profile 似然函数 (式 (3.13)) 利用得分方法来构造检验统计量, 可得如下结果.

定理 3.5 设 $D = [\partial W(R_i,\theta)/\partial\theta^{\mathrm{T}}]|_{\theta=\theta_0} = (D_{ij})_{n\times k}$, 记 $U = (\hat{\varepsilon}_i^2)_{n\times 1}$, 其中 $\hat{\varepsilon}_i = \widetilde{y}_i - \widetilde{\boldsymbol{z}}_i^{\mathrm{T}}\hat{\boldsymbol{\beta}}_{\mathrm{PM}}$, $\overline{D} = (\boldsymbol{I}_n, -\boldsymbol{1}_n, \boldsymbol{1}_n^{\mathrm{T}})D$, 则基于 profile 似然函数 (3.13), 假设检验 (3.11) 的 profile 得分检验统计量可表示为

$$T = U^{\mathrm{T}}\overline{D}(\overline{D}^{\mathrm{T}}\overline{D})^{-1}\overline{D}^{\mathrm{T}}U/2\hat{\sigma}^4_{\mathrm{PM}},$$

特别地, 若 $k=1$, 令 $d_i = [\partial W(R_i,\theta)/\partial\theta^{\mathrm{T}}]|_{\theta=\theta_0}$, $\overline{d} = \sum_{i=1}^n d_i/n$, $s_i = d_i - \overline{d}$, 则有

$$T = \left(\sum_{i=1}^n s_i\hat{\varepsilon}_i^2\right)^2 \Bigg/ \left(2\hat{\sigma}^4_{\mathrm{PM}}\sum_{i=1}^n s_i^2\right).$$

为了克服检验统计量 T 依赖于假设 $\boldsymbol{\varepsilon} \sim N(\boldsymbol{0}, \sigma^2\boldsymbol{I}_n)$ 这一局限性, Koenker(1981) 基于线性模型提出了学生化 Score 统计量, 即用下面的量来代替 $2\hat{\sigma}^4_{\mathrm{PM}}$

$$\psi = \sum_{i=1}^n (\hat{\varepsilon}_i^2 - \hat{\sigma}^2_{\mathrm{PM}})^2/n.$$

所以可定义相应的稳健 profile 得分检验统计量为

$$T^* = U^{\mathrm{T}}\overline{D}(\overline{D}^{\mathrm{T}}\overline{D})^{-1}\overline{D}^{\mathrm{T}}U/\psi.$$

由部分线性变系数模型 profile 最小二乘估计的性质可知, 如果原假设 H_0 成立, 在一些常规条件下 T 的渐近分布为 χ^2_k, 其中 χ^2_k 表示自由度为 k 的 χ^2 分布.

注 3.5 上面得分检验统计量等价于基于 profile 似然函数构造的 Lagrange 乘子检验统计量, 由于 $\hat{\boldsymbol{\beta}}_{\mathrm{PM}}$ 与 $\hat{\sigma}^2_{\mathrm{PM}}$ 不是极大似然估计, 所以检验统计量 T 也可从另一方面认为是 Neyman (1959) 提出的 $C(\alpha)$ 检验 (拟 Lagrange 乘子检验) 在半参数模型中的一个推广.

注 3.6 上面对于模型的估计应用的是基于局部线性方法的 profile 最小二乘估计, 如果利用其他光滑方法, 比如级数方法 (包括 B 样条与幂基函数法), 最后所得检验统计量形式与 T 相同, 只是拟合残差是基于相应的估计方法. 这也说明本文提出方法对于半参数模型具有很强的适用性. 而且上面的方法很明显对部分线性模型同样适用.

3.4 关于一类部分线性模型异方差检验的一个注记

You 和 Chen (2005) 针对一类部分线性模型研究了异方差检验问题, 所用检验方法是 Dette 和 Munk (1999) 针对一类非参数回归模型所提出的检验方法的直接推广. 下面首先介绍 Dette 和 Munk (1999) 的方法.

考虑如下的非参数回归模型

$$y_i = g(t_i) + \sigma(t_i)\varepsilon_i, \quad i = 1, 2, \cdots, n, \tag{3.17}$$

其中 $0 \leqslant t_1 < t_2 < \cdots < t_n \leqslant 1$ 为固定设计点列, $g(\cdot)$ 为均值函数, $\sigma^2(\cdot)$ 为未知方差函数, 误差 ε_i 假定形成三角形的矩阵, 而且行间互相独立的均值为零方差为 1 的随机变量. 假定设计空间在 [0,1] 上是紧的, 而且设计序列形成一个渐近正规的序列, 即

$$\max_{i=1,2,\cdots,n}\left|\int_0^{t_i} h(t)\mathrm{d}t - \frac{i-1}{n-1}\right| = o(n^{-3/2}),$$

这里 h 表示 [0,1] 区间的一个正密度函数, 而且它为 $\gamma > 0$ 阶 Lipschitz 连续, 即 $h \in \mathrm{Lip}_\gamma[0,1]$. 这是我们的第一个假定 (B.1), 其实这包含了任何光滑的正密度, 特别是 [0,1] 上均匀设计的常数密度. 感兴趣的异方差的检验问题可记为

$$H_0: \quad \sigma^2(\cdot) \equiv \sigma^2 \qquad \text{VS} \qquad H_1: \quad \sigma^2(\cdot) \neq \sigma^2.$$

自然地, 选择下面的统计量度量方差齐性的程度,

$$T_n^2 = \frac{1}{n}\sum_{i=1}^n \left\{\sigma^2(t_i) - \frac{1}{n}\sum_{j=1}^n \sigma^2(t_j)\right\}^2.$$

显然对于方差齐性的检验问题等价于检验

$$H_0: \quad T_n^2 = 0 \qquad VS \qquad H_1: \quad T_n^2 \neq 0.$$

为了找出关于异方差的度量, 用方差函数最好的 L^2 近似:

$$M_\sigma^2 = \min_{a\in\mathbf{R}^+}\left[\int_0^1 \{\sigma^2(t) - a\}^2 h(t)\mathrm{d}t\right] = \int_0^1 \sigma^4(t)h(t)\mathrm{d}t - \left\{\int_0^1 \sigma^2(t)h(t)\mathrm{d}t\right\}^2.$$

基于前面的假设 (B1), 有 $\lim\limits_{n\to\infty} T_n^2 = M_\sigma^2$. 定义其估计

$$\hat{M}_n^2 = \frac{1}{4(n-3)}\sum_{j=2}^{n-2} R_j^2 R_{j+2}^2 - \left\{\frac{1}{2(n-2)}\sum_{j=2}^n R_j^2\right\}^2, \tag{3.18}$$

其中伪残差 $R_j = y_j - y_{j-1}(j = 2, \cdots, n)$. 为了证明 $\hat{M}_n^2$ 为 M_σ^2 的渐近无偏估计, 给出第二个假定 (B2):

$$m_3, m_4, g, \sigma^2 \in \mathrm{Lip}_\gamma[0,1], \quad \gamma > 0$$

这里 $m_i(t) = E[\varepsilon(t)^i], i = 3, 4$. 对于某个 $\gamma > 0$, 如果上面两个假定 (B1), 假定 (B2) 满足, 则 $\hat{M}_n^2$ 为渐近无偏的, 即

$$E[\hat{M}_n^2] = T_n^2 + O(\frac{1}{n^\gamma}) = M_\sigma^2 + O\left(\frac{1}{n^\gamma}\right).$$

上面的结论表明可以基于 $\hat{M}_n^2$ 来检验异方差. 最后 Dette 和 Munk (1998) 在两个假定下证明了下面的结论

$$4n^{1/2}(\hat{M}_n^2 - M_\sigma^2) \to N(0, v_\sigma^2),$$

渐近方差为

$$\begin{aligned} v_\sigma^2 = \int_0^1 &\{[6m_4^2(t) + 4m_4(t) + 6]\sigma^8(t) + 64m_4(t)\sigma^4(t)(\sigma^2(t) - \overline{\sigma}^2) \\ &- 8m_3^2(t)\sigma^6(t)(\sigma^2(t) - \overline{\sigma}^2)\}h(t)\mathrm{d}t, \end{aligned}$$

其中

$$\overline{\sigma}^2 = \int_0^1 \sigma^2(t)h(t)\mathrm{d}t.$$

定义

$$\hat{A}_{4,n} = \frac{1}{2(n-1)}\sum_{j=2}^n R_j^4, \quad \hat{S}_{2,n}^2 = \frac{1}{2(n-1)}\sum_{j=2}^n R_j^2, \tag{3.19}$$

与

$$\hat{T}_n^2 = \hat{M}_n^2 + \frac{1}{n-1}(\hat{A}_{4,n} - \hat{S}_{2,n}^2),$$

那么检验统计量为

$$4n^{1/2}\hat{T}_n^2/\hat{v}_n,$$

其中 $\hat{v}_n^2$ 为 v_σ^2 的估计

$$\begin{aligned} \hat{v}_n^2 = &\frac{3}{2(n-3)}\sum_{j=2}^{n-2} R_j^4 R_{j+2}^4 - \frac{4}{n-5}\sum_{j=2}^{n-4} R_j^4 R_{j+2}^2 R_{j+4}^2 + \frac{3}{16(n-1)^8}\left(\sum_{j=2}^n R_j^2\right)^8 \\ &+ \frac{1}{9(n-7)}\sum_{j=3}^{n-5}(R_j - R_{j-1})^3(R_{j+3} - R_{j+2})^3 R_{j+5}^2. \end{aligned} \tag{3.20}$$

如果显著水平为 α, 那么如果

$$4n^{1/2}\hat{T}_n^2/\hat{v}_n > z_{1-\alpha},$$

则拒绝原假设. 其中 $z_{1-\alpha}$ 标准正态分布的 $1-\alpha$ 分位点.

基于上面的思想, You 和 Chen(2005) 研究了下面的异方差部分线性模型

$$y_i = \boldsymbol{x}_i^{\mathrm{T}}\boldsymbol{\beta} + g(t_i) + \sigma(t_i)\varepsilon_i, \quad i = 1, 2, \cdots, n, \tag{3.21}$$

其中 y_i 为因变量观测值, $\boldsymbol{\beta} = (\beta_1, \beta_2, \cdots, \beta_p)^{\mathrm{T}}$ 为未知参数向量, $g(\cdot)$ 与 $\sigma(\cdot)$ 为未知光滑函数, $\boldsymbol{x}_i = (x_{i2}, x_{i1}, \cdots, x_{ip})^{\mathrm{T}}$ 和 t_i 为固定设计点列, 其他关于 t 以及 ε 假定同上面的假定 (B1).

You 和 Chen (2005) 将上面的部分线性模型转化为如下的拟非参数回归模型

$$y_i^* = y_i - \boldsymbol{x}_i^{\mathrm{T}}\hat{\boldsymbol{\beta}} = g(t_i) + \sigma(t_i)\varepsilon_i, \quad i = 1, 2, \cdots, n,$$

从而对于异方差的检验问题

$$H_0 : \sigma^2(\cdot) = \sigma^2 \quad \text{VS} \quad H_1 : \sigma^2(\cdot) \neq \sigma^2.$$

证明了可直接应用 Dette 和 Munk(1998) 的检验方法. 其中 $\hat{\boldsymbol{\beta}}$ 为 $\boldsymbol{\beta}$ 的任一 $\sqrt{n}$ 相合估计, 文献中有很多种方法可以用来估计模型而得到 $\boldsymbol{\beta}$ 的 $\sqrt{n}$ 相合估计, 具体的方法可参考 Spenkman (1988), Chen (1988) 等. You 和 Chen (2005) 利用了常见的 Speckman (1988) 提出的基于核方法的 profile 最小二乘估计方法.

You 和 Chen (2005) 关于部分线性模型异方差的检验问题的具体操作步骤如下:

(1) 对部分线性模型 (3.21) 采用一种估计方法, 得到 $\boldsymbol{\beta}$ 的 $\sqrt{n}$ 相合估计 $\hat{\boldsymbol{\beta}}$,

(2) 定义 $y_i^* = y_i - \boldsymbol{x}_i^{\mathrm{T}}\hat{\boldsymbol{\beta}}$ $(i = 1, 2, \cdots, n)$ 与 $R_j^* = y_j^* - y_{j-1}^*$ $(j = 2, 3, \cdots, n)$,

(3) 计算检验统计量 $4n^{-1/2}\hat{T}_n^{*2}/\hat{v}_n^*$, 该统计量与 Dette 和 Munk(1998) 检验统计量的形式一样, 只是将式 (3.18)~ 式 (3.20) 中的 R_j 替换为 R_j^*,

(4) 取显著水平为 α, 那么若

$$4n^{-1/2}\hat{T}_n^{*2}/\hat{v}_n^* > Z_{1-\alpha},$$

就拒绝原假设 H_0, 即认为模型 (3.21) 不满足方差齐性. 其中 $Z_{1-\alpha}$ 为标准正态分布的 $1-\alpha$ 分位数.

需要说明的是如果利用基于核方法的 profile 最小二乘估计来拟合部分线性模型, 必须涉及窗宽的选择问题, You 和 Chen (2005) 在模拟中用 Cross-Validation 方法来选择窗宽. Dette 和 Munk (1998) 检验统计量的构造是基于差分方法, 该方法

的最大特点就是利用差分的思想消除非参数的影响, 而不用任何光滑方法. 对于部分线性模型 (3.21), 可以利用该方法得到 $\boldsymbol{\beta}$ 的估计, 该估计不但是 $\sqrt{n}$ 相合的, 而且具体计算中不涉及窗宽的选择, 从而只利用该方法就能构造相关的检验统计量, 而不是像 You 和 Chen (2005) 还要应用其他光滑方法从而存在窗宽的选择问题, 增大了计算量, 忽视了部分线性模型差分估计的简便性. 下面给出具体的做法.

对部分线性模型 (3.21) 进行一阶差分可得

$$\begin{aligned} y_i - y_{i-1} &= (\boldsymbol{x}_i - \boldsymbol{x}_{i-1})^{\mathrm{T}}\boldsymbol{\beta} + (g(t_i) - g(t_{i-1})) + \varepsilon_i - \varepsilon_{i-1} \\ &\approx (\boldsymbol{x}_i - \boldsymbol{x}_{i-1})^{\mathrm{T}}\boldsymbol{\beta} + (\varepsilon_i - \varepsilon_{i-1}). \end{aligned}$$

从而对上面的线性回归模型利用最小二乘法可得

$$\hat{\boldsymbol{\beta}}_{\text{diff}} = \left[\sum_{i=2}^{n}(\boldsymbol{x}_i - \boldsymbol{x}_{i-1})(\boldsymbol{x}_i - \boldsymbol{x}_{i-1})^{\mathrm{T}}\right]^{-1} \sum_{i=2}^{n}(\boldsymbol{x}_i - \boldsymbol{x}_{i-1})(y_i - y_{i-1}).$$

那么有基于差分估计的伪残差

$$R_i^{**} = (y_i - y_{i-1}) - (\boldsymbol{x}_i - \boldsymbol{x}_{i-1})^{\mathrm{T}}\hat{\boldsymbol{\beta}}_{\text{diff}}, \quad i = 2, 3, \cdots, n, \tag{3.22}$$

从而基于该伪残差 R_i^{**} 计算检验统计量 $4n^{-1/2}\hat{T}_n^{*2}/\hat{v}_n^{*}$. 显然我们的方法与 You 和 Chen (2005) 相比非常简便.

3.5 定理的证明

下面给出定理 3.1 到定理 3.5 的证明。首先给出一些条件. 为了本章内容的独立性, 下面给出的条件以及引理与第 2 章有所重复, 但仍然列出.

(A.1) 随机变量 U 具有有界支撑 $\boldsymbol{\Omega}$, 其密度函数 $f(\cdot)$ 在其支撑上满足 Lipschitz 连续, 且不为 0.

(A.2) 对于任一 $U \in \boldsymbol{\Omega}$, 矩阵 $E(\boldsymbol{X}\boldsymbol{X}^{\mathrm{T}}|U)_{k\times K}$ 为非奇异, $E(\boldsymbol{X}\boldsymbol{X}^{\mathrm{T}}|U)_{k\times K}$, $E(\boldsymbol{X}\boldsymbol{X}^{\mathrm{T}}|U)_{k\times K}^{-1}$ 和 $E(\mathbf{X}\mathbf{Z}^{\mathrm{T}}|U)_{k\times K}$ 都是 Lipschitz 连续的.

(A.3) 存在 $s > 2$, 使得 $E\|\boldsymbol{X}\|^{2s} < \infty$ 和 $E\|\boldsymbol{Z}\|^{2s} < \infty$, 对于 $\varepsilon < 2 - s^{-1}$, 使得 $n^{2\varepsilon-1}h \to \infty$.

(A.4) $\alpha(\cdot)$ 二阶连续可导.

(A.5) 函数 $K(\cdot)$ 为对称密度函数, 具有紧支撑.

(A.6) $nh^8 \to 0$ 和 $nh^2/(\log n)^2 \to \infty$.

注 上述条件是 Fan 和 Huang (2005) 证明定理所附加的条件, 这些条件只是为了证明的方便, 并非最弱条件.

下面记$\mu_i = \int_0^\infty \mu^i K(\mu)\mathrm{d}\mu, \nu_i = \int_0^\infty \mu^i K^2(\mu)\mathrm{d}\mu$, $c_n = \left\{\frac{\log(1/h)}{nh}\right\}^{1/2}$. 下面给出证明所需要的几个引理.

引理 3.1 令 $(\boldsymbol{X}_1,\boldsymbol{Y}_1),\cdots,(\boldsymbol{X}_n,\boldsymbol{Y}_n)$ 为独立同分布 (iid) 的随机序列, 其中 $\boldsymbol{Y}_i, i=1,2,\cdots,n$ 为一元随机变量, 进一步假定 $E|y|^s<\infty$ 与 $\sup_x \int |y|^s f(x,y)\mathrm{d}y<\infty$, 其中 f 表示 $(\boldsymbol{X},\boldsymbol{Y})$ 的联合密度. 令 K 为一有界正函数, 并有有界支撑且满足 Lipschitz 条件, 则有

$$\sup_x \left|\frac{1}{n}\sum_{i=1}^n [K_h(\boldsymbol{X}_i - x)\boldsymbol{Y}_i - E\left(K_h(\boldsymbol{X}_i - x)\boldsymbol{Y}_i\right)]\right| = O_p\left(\left\{\frac{\log(1/h)}{nh}\right\}^{1/2}\right),$$

其中对于 $\varepsilon < 1 - s^{-1}$ 有 $n^{2\varepsilon-1}h \to \infty$.

该引理由 Mack 和 Silverman (1982) 即可证得.

引理 3.2 假设上面的条件 (A.1)~ 条件 (A.6) 成立, 有

$$n^{-1}\overline{\boldsymbol{Z}}^{\mathrm{T}}\overline{\boldsymbol{Z}} \to \boldsymbol{\varXi} = Ee_ie_i^{\mathrm{T}},$$

其中 $\boldsymbol{e}_i = \boldsymbol{z}_i - \Phi^{\mathrm{T}}(u_i)\Gamma^{-1}(u_i)\boldsymbol{x}_i$.

引理 3.3 假设上面的条件 (A.1)~ 条件 (A.6) 成立, 有

$$n^{-1}\overline{\boldsymbol{Z}}^{\mathrm{T}}(\boldsymbol{I}-\boldsymbol{S})\boldsymbol{M} = O_p(c_n^2).$$

注 引理 3.2 即为 Fan 和 Huang (2005) 的定理 7.2, 引理 3.3 为 Fan 和 Huang (2005) 的引理 7.4.

定理 3.1 的证明 由

$$\begin{aligned}\hat{\boldsymbol{\beta}}_{\mathrm{PL}} &= [\overline{\boldsymbol{Z}}^{\mathrm{T}}\overline{\boldsymbol{Z}}]^{-1}\overline{\boldsymbol{Z}}^{\mathrm{T}}(\boldsymbol{I}-\boldsymbol{S})\boldsymbol{Y}\\ &= [\overline{\boldsymbol{Z}}^{\mathrm{T}}\overline{\boldsymbol{Z}}]^{-1}\overline{\boldsymbol{Z}}^{\mathrm{T}}(\boldsymbol{I}-\boldsymbol{S})(\boldsymbol{Z}\boldsymbol{\beta}+\boldsymbol{M}+\boldsymbol{\varepsilon})\\ &= \boldsymbol{\beta} + [\overline{\boldsymbol{Z}}^{\mathrm{T}}\overline{\boldsymbol{Z}}]^{-1}\overline{\boldsymbol{Z}}^{\mathrm{T}}(\boldsymbol{I}-\boldsymbol{S})\boldsymbol{M} + [\overline{\boldsymbol{Z}}^{\mathrm{T}}\overline{\boldsymbol{Z}}]^{-1}\overline{\boldsymbol{Z}}^{\mathrm{T}}(\boldsymbol{I}-\mathrm{S})\boldsymbol{\varepsilon},\end{aligned}$$

可得

$$\sqrt{n}(\hat{\boldsymbol{\beta}}_{\mathrm{PL}}-\boldsymbol{\beta}) = \sqrt{n}[\overline{\boldsymbol{Z}}^{\mathrm{T}}\overline{\boldsymbol{Z}}]^{-1}\overline{\boldsymbol{Z}}^{\mathrm{T}}(\boldsymbol{I}-\boldsymbol{S})\boldsymbol{M} + \sqrt{n}[\overline{\boldsymbol{Z}}^{\mathrm{T}}\overline{\boldsymbol{Z}}]^{-1}\overline{\boldsymbol{Z}}^{\mathrm{T}}(\boldsymbol{I}-\boldsymbol{S})\boldsymbol{\varepsilon}.$$

由引理 3.2 和引理 3.3 可得

$$\sqrt{n}[\overline{\boldsymbol{Z}}^{\mathrm{T}}\overline{\boldsymbol{Z}}]^{-1}\overline{\boldsymbol{Z}}^{\mathrm{T}}(\boldsymbol{I}-\boldsymbol{S})\boldsymbol{M} = O_p(\sqrt{n}c_n^2) = o_p(1),$$

与

$$\sqrt{n}[\overline{\boldsymbol{Z}}^{\mathrm{T}}\overline{\boldsymbol{Z}}]^{-1}\overline{\boldsymbol{Z}}^{\mathrm{T}}(\boldsymbol{I}-\boldsymbol{S})\boldsymbol{\varepsilon} = n^{-1/2}\boldsymbol{\varXi}^{-1}\overline{\boldsymbol{Z}}^{\mathrm{T}}(\boldsymbol{I}-\boldsymbol{S})\boldsymbol{\varepsilon}\{1+o_p(1)\}.$$

由引理 3.1 计算可得

$$\overline{\boldsymbol{Z}}^{\mathrm{T}}(\boldsymbol{I}-\boldsymbol{S})\boldsymbol{\varepsilon}=\sum_{i=1}^{n}\{\boldsymbol{z}_i-\boldsymbol{\Phi}^{\mathrm{T}}(u_i)\Gamma^{-1}(u_i)\boldsymbol{x}_i\}\varepsilon_i\{1+o_p(1)\}=\sum_{i=1}^{n}\boldsymbol{e}_i\varepsilon_i\{1+o_p(1)\}.$$

对于 $\boldsymbol{e}_i\varepsilon_i$ 有

$$\begin{aligned}&\boldsymbol{E}(\boldsymbol{e}_i\varepsilon_i)=0,\\&\mathrm{Var}(\boldsymbol{e}_i\varepsilon_i)=E[\boldsymbol{e}_i\boldsymbol{e}_i^{\mathrm{T}}\varepsilon_i^2]=E[\boldsymbol{e}_i\boldsymbol{e}_i^{\mathrm{T}}E[\varepsilon_i^2|\boldsymbol{x}_i,\boldsymbol{z}_i,u_i]]=E[\boldsymbol{e}_i\boldsymbol{e}_i^{\mathrm{T}}\sigma^2(\boldsymbol{x}_i,\boldsymbol{z}_i,u_i)]=\boldsymbol{\Omega}.\end{aligned}$$

由中心极限定理可得

$$n^{-1/2}\overline{\boldsymbol{Z}}^{\mathrm{T}}(\boldsymbol{I}-\boldsymbol{S})\boldsymbol{\varepsilon}\to N(\mathbf{0},\boldsymbol{\Omega}).$$

从而由 Slutsky 定理可得

$$\sqrt{n}(\hat{\boldsymbol{\beta}}_{\mathrm{PL}}-\boldsymbol{\beta})\to N(\mathbf{0},\boldsymbol{\Xi}^{-1}\boldsymbol{\Omega}\boldsymbol{\Xi}^{-1}).$$

证毕.

定理 3.2 的证明　首先有

$$\hat{\varepsilon}_i=\overline{y}_i-\overline{\boldsymbol{z}}_i^{\mathrm{T}}\hat{\boldsymbol{\beta}}_{\mathrm{PL}}=\varepsilon_i+(\boldsymbol{\alpha}^{\mathrm{T}}(u_i)\boldsymbol{x}_i-\boldsymbol{S}_i^{\mathrm{T}}\boldsymbol{M})+\overline{\boldsymbol{z}}_i^{\mathrm{T}}(\boldsymbol{\beta}-\hat{\boldsymbol{\beta}}_{\mathrm{PL}})-\boldsymbol{S}_i^{\mathrm{T}}\boldsymbol{\varepsilon}.$$

由引理 3.1 可证得

$$\boldsymbol{\alpha}^{\mathrm{T}}(u_i)\boldsymbol{x}_i-\boldsymbol{S}_i^{\mathrm{T}}\boldsymbol{M}=O_p(c_n),\quad \boldsymbol{S}_i^{\mathrm{T}}\boldsymbol{\varepsilon}=O_p(c_n).$$

再由定理 3.1 可知

$$\boldsymbol{\beta}-\hat{\boldsymbol{\beta}}_{\mathrm{PL}}\to\mathbf{0}.$$

利用上面的结论可得

$$\frac{1}{n}\sum_{i=1}^{n}\widetilde{\boldsymbol{z}}_i\widetilde{\boldsymbol{z}}_i^{\mathrm{T}}\hat{\varepsilon}_i^2=\frac{1}{n}\sum_{i=1}^{n}\widetilde{\boldsymbol{z}}_i\widetilde{\boldsymbol{z}}_i^{\mathrm{T}}\varepsilon_i^2+o_p(1).$$

由大数定律得

$$\frac{1}{n}\sum_{i=1}^{n}\widetilde{\boldsymbol{z}}_i\widetilde{\boldsymbol{z}}_i^{\mathrm{T}}\varepsilon_i^2\to\boldsymbol{\Omega},$$

再由引理 3.2 从而可得

$$\left(\frac{1}{n}\sum_{i=1}^{n}\widetilde{\boldsymbol{z}}_i\widetilde{\boldsymbol{z}}_i^{\mathrm{T}}\right)^{-1}\left(\frac{1}{n}\sum_{i=1}^{n}\widetilde{\boldsymbol{z}}_i\widetilde{\boldsymbol{z}}_i^{\mathrm{T}}\hat{\varepsilon}_i^2\right)\left(\frac{1}{n}\sum_{i=1}^{n}\widetilde{\boldsymbol{z}}_i\widetilde{\boldsymbol{z}}_i^{\mathrm{T}}\right)^{-1}\to\boldsymbol{\Xi}^{-1}\boldsymbol{\Omega}\boldsymbol{\Xi}^{-1}.$$

证毕.

定理 3.3 的证明完全类似于定理 3.1 的证明, 故在此省略.

定理 3.4 可由定理 3.1 直接得到, 故在此省略.

定理 3.5 的证明 下面基于 profile 似然函数

$$L(\hat{\boldsymbol{\alpha}}(\cdot;\boldsymbol{\beta}),\boldsymbol{\beta},\sigma^2,\boldsymbol{\theta})C-\frac{n}{2}\log\sigma^2-\frac{1}{2}\sum_{i=1}^{n}\log W(R_i,\theta)-\sum_{i=1}^{n}\frac{(\widetilde{y}_i-\boldsymbol{\beta}^{\mathrm{T}}\widetilde{\boldsymbol{z}}_i)^2}{2\sigma^2W(R_i,\theta)}$$

来构造检验统计量, 将 $L(\hat{\boldsymbol{\alpha}}(\cdot;\boldsymbol{\beta}),\boldsymbol{\beta},\sigma^2,\boldsymbol{\theta})$ 看作只包含未知参数 $\boldsymbol{\beta},\sigma^2,\boldsymbol{\theta}$ 而与无穷维参数 $\boldsymbol{\alpha}(\cdot)$ 无关. 由 Breusch 和 Pagan (1979) 可知, 针对异方差检验问题 (3.11) 的 Score 检验统计量为

$$\begin{aligned}C(\alpha)=&(\partial L/\partial\boldsymbol{\theta}-I_{\boldsymbol{\theta\phi}}I_{\boldsymbol{\phi\phi}^{\mathrm{T}}}^{-1}\partial L/\partial\boldsymbol{\phi})^{\mathrm{T}}(I_{\boldsymbol{\theta\theta}^{\mathrm{T}}}-I_{\boldsymbol{\theta\phi}}I_{\boldsymbol{\phi\phi}^{\mathrm{T}}}^{-1}I_{\boldsymbol{\theta\phi}}^{\mathrm{T}})^{-1}\\&\times(\partial L/\partial\boldsymbol{\theta}-I_{\boldsymbol{\theta\phi}}I_{\boldsymbol{\phi\phi}^{\mathrm{T}}}^{-1}\partial L/\partial\boldsymbol{\phi})|_{(\boldsymbol{\theta},\boldsymbol{\phi})=(\boldsymbol{\theta}_0,\boldsymbol{\phi}_0)},\end{aligned}$$

其中记冗余参数 $\boldsymbol{\phi}=(\sigma^2,\boldsymbol{\beta}^{\mathrm{T}})^{\mathrm{T}}$, I_{ij} 为参数 $(\boldsymbol{\theta},\boldsymbol{\phi})$ 对应的 Fisher 信息阵 $I(\boldsymbol{\theta};\boldsymbol{\phi})$ 的元素. $\boldsymbol{\phi}_0$ 即为原假设成立时 $\sigma^2,\boldsymbol{\beta}$ 对应的估计值 $\hat{\sigma}^2_{\mathrm{PM}}$ 与 $\hat{\boldsymbol{\beta}}_{\mathrm{PM}}$. 基于下面的式子

$$\begin{aligned}
&\partial L(\hat{\boldsymbol{\alpha}}(\cdot;\boldsymbol{\beta}),\boldsymbol{\beta},\sigma^2,\theta)/\partial\boldsymbol{\theta}=-\frac{1}{2}\sum_{i=1}^{n}\frac{D_i}{W_i}+\frac{1}{2\sigma^2}\sum_{i=1}^{n}\frac{D_i(\widetilde{y}_i-\widetilde{\boldsymbol{z}}_i^{\mathrm{T}}\boldsymbol{\beta})^2}{W_i^2},\\
&\partial L(\hat{\boldsymbol{\alpha}}(\cdot;\boldsymbol{\beta}),\boldsymbol{\beta},\sigma^2,\theta)/\partial\boldsymbol{\beta}=\sum_{i=1}^{n}\frac{\widetilde{\boldsymbol{z}}_i(\widetilde{y}_i-\widetilde{\boldsymbol{z}}_i^{\mathrm{T}}\boldsymbol{\beta})}{\sigma^2W_i},\\
&\partial L(\hat{\boldsymbol{\alpha}}(\cdot;\boldsymbol{\beta}),\boldsymbol{\beta},\sigma^2,\theta)/\partial\sigma^2=-\frac{n}{2\sigma^2}+\frac{1}{2\sigma^4}\sum_{i=1}^{n}\frac{(\widetilde{y}_i-\widetilde{\boldsymbol{z}}_i^{\mathrm{T}}\boldsymbol{\beta})^2}{W_i},\\
&\partial^2L(\hat{\boldsymbol{\alpha}}(\cdot;\boldsymbol{\beta}),\boldsymbol{\beta},\sigma^2,\theta)/\partial\boldsymbol{\beta}\partial\boldsymbol{\theta}^{\mathrm{T}}=-\sum_{i=1}^{n}\frac{D_i\widetilde{\boldsymbol{z}}_i(\widetilde{y}_i-\widetilde{\boldsymbol{z}}_i^{\mathrm{T}}\boldsymbol{\beta})}{\sigma^2W_i^2},\\
&\partial^2L(\hat{\boldsymbol{\alpha}}(\cdot;\boldsymbol{\beta}),\boldsymbol{\beta},\sigma^2,\theta)/\partial\boldsymbol{\beta}\partial\sigma^2=-\frac{1}{2\sigma^4}\sum_{i=1}^{n}\frac{\widetilde{\boldsymbol{z}}_i(\widetilde{y}_i-\widetilde{\boldsymbol{z}}_i^{\mathrm{T}}\boldsymbol{\beta})}{W_i},\\
&\partial^2L(\hat{\boldsymbol{\alpha}}(\cdot;\boldsymbol{\beta}),\boldsymbol{\beta},\sigma^2,\theta)/\partial\boldsymbol{\theta}\partial\sigma^2=-\frac{1}{2\sigma^4}\sum_{i=1}^{n}\frac{D_i(\widetilde{y}_i-\widetilde{\boldsymbol{z}}_i^{\mathrm{T}}\boldsymbol{\beta})^2}{W_i^2},\\
&\partial^2L(\hat{\boldsymbol{\alpha}}(\cdot;\boldsymbol{\beta}),\boldsymbol{\beta},\sigma^2,\theta)/\partial\sigma^2\partial\sigma^2=\frac{n}{2\sigma^4}-\frac{1}{2\sigma^6}\sum_{i=1}^{n}\frac{(\widetilde{y}_i-\widetilde{\boldsymbol{z}}_i^{\mathrm{T}}\boldsymbol{\beta})^2}{W_i},\\
&\partial^2L(\hat{\boldsymbol{\alpha}}(\cdot;\boldsymbol{\beta}),\boldsymbol{\beta},\sigma^2,\theta)/\partial\boldsymbol{\beta}\partial\boldsymbol{\beta}^{\mathrm{T}}=-\frac{1}{\sigma^2}\sum_{i=1}^{n}\widetilde{\boldsymbol{z}}_i\widetilde{\boldsymbol{z}}_i^{\mathrm{T}},\\
&\partial^2L(\hat{\boldsymbol{\alpha}}(\cdot;\boldsymbol{\beta}),\boldsymbol{\beta},\sigma^2,\theta)/\partial\boldsymbol{\theta}\partial\boldsymbol{\theta}^{\mathrm{T}}=-\frac{1}{2}\sum_{i=1}^{n}\frac{D_iD_i^{\mathrm{T}}}{W_i^2}-\frac{1}{2\sigma^2}\sum_{i=1}^{n}\frac{D_iD_i^{\mathrm{T}}(\widetilde{y}_i-\widetilde{\boldsymbol{z}}_i^{\mathrm{T}}\boldsymbol{\beta})^2}{W_i^3},
\end{aligned}$$

求出参数对应的 Fisher 信息阵, 并在 $(\boldsymbol{\theta},\boldsymbol{\phi})=(\boldsymbol{\theta}_0,\boldsymbol{\phi}_0)$ 处取值可得

$$I(\boldsymbol{\theta}=\boldsymbol{\theta}_0;\sigma^2=\hat{\sigma}^2_{\mathrm{PM}};\boldsymbol{\beta}=\hat{\boldsymbol{\beta}}_{\mathrm{PM}})=\begin{pmatrix}\frac{1}{2}D^{\mathrm{T}}D & \frac{1}{2\hat{\sigma}^2_{\mathrm{PM}}}D^{\mathrm{T}}\mathbf{1} & \mathbf{0}\\ \frac{1}{2\hat{\sigma}^2_{\mathrm{PM}}}\mathbf{1}^{\mathrm{T}}D & \frac{1}{2\hat{\sigma}^4_{\mathrm{PM}}} & \mathbf{0}\\ \mathbf{0} & \mathbf{0} & \overline{\boldsymbol{Z}}^{\mathrm{T}}\overline{\boldsymbol{Z}}/\hat{\sigma}^2_{\mathrm{PM}},\end{pmatrix}.$$

以及

$$\partial L/\partial\boldsymbol{\theta}|_{\boldsymbol{\theta}=\boldsymbol{\theta}_0}=\frac{1}{2\hat{\sigma}^2_{\mathrm{PM}}}D^{\mathrm{T}}(R-\hat{\sigma}^2_{\mathrm{PM}}\mathbf{1}),$$

与

$$\partial L/\partial\boldsymbol{\phi}|_{\boldsymbol{\theta}=\boldsymbol{\phi}_0}=\mathbf{0}.$$

将上式代入 $C(\alpha)$ 经过计算得

$$C(\alpha)=U^{\mathrm{T}}\overline{D}(\overline{D}^{\mathrm{T}}\overline{D})^{-1}\overline{D}^{\mathrm{T}}U/2\hat{\sigma}^4_{\mathrm{PM}}.$$

证毕.

第 4 章　部分线性变系数变量含误差模型的约束估计与检验

利用回归模型进行实际数据分析时, 由于各种原因, 往往需要对模型系数附加有约束条件, 回归模型在含有约束条件下的推断问题是回归分析的一个重要分支, 然而大部分已有结果都是基于普通的线性回归模型得到的. 此外, 回归模型的自变量有时不能精确观测, 由此发展起来的变量含误差模型 (测量误差模型) 在统计学和计量经济学领域都得到了广泛的关注和深入的研究. 本章重点研究一类半参数变量含误差模型在含有约束条件下的统计推断问题, 主要内容来自于文献 Wei (2012).

4.1　引　　言

近三十年以来, 变量含误差模型在统计学和计量经济学领域都得到了深入研究. 然而, 目前现有理论大都是关于线性和非线性变量含误差模型的结果, 具体内容可参考 Fuller (1987), Cheng 和 Ness (1999) 以及 Carroll 等 (2006). 随着计算机计算能力的发展, 近年来非参数和半参数变量含误差模型也得到越来越多的研究. 本章将研究如下的部分线性变系数变量含误差模型.

$$\begin{cases} \boldsymbol{Y} = \boldsymbol{X}^{\mathrm{T}}\boldsymbol{\beta} + \boldsymbol{Z}^{\mathrm{T}}\boldsymbol{\alpha}(U) + \varepsilon, \\ \boldsymbol{V} = \boldsymbol{X} + \eta, \end{cases} \tag{4.1}$$

其中 Y 是因变量, Z, X 和 U 为自变量. 同时为了表述方便, 假设 U 为一元变量, $\boldsymbol{\beta} = (\beta_1, \beta_2, \cdots, \beta_p)^{\mathrm{T}}$ 为 p 维未知参数, $\boldsymbol{\alpha}(\cdot) = (\alpha_1(\cdot), \alpha_1(\cdot), \cdots, \alpha_q(\cdot))^{\mathrm{T}}$ 是 q 维未知函数系数, ε 为模型随机误差, 满足 $E[\varepsilon|X, Z, U] = 0$, 测量误差 η 为零均值, 协方差矩阵为 $\boldsymbol{\Sigma}_\eta$, 并且与 (Y, X, Z, U) 独立. 文中将假定 $\boldsymbol{\Sigma}_\eta$ 已知; 否则可以利用 V 的重复观测去估计它.

显然, 当 $q = 1$ 和 $Z = 1$ 时, 模型 (4.1) 即为部分线性变量含误差模型, 该模型已经得到了广泛的研究, 具体可参考 Liang 等 (1999), Liang 等 (2007), Cui 和 Li (1998), Cui 和 Kong (2005), Wang (1999). 对于模型 (4.1), You 和 Chen (2006) 提出了一类校正 profile 最小二乘估计方法. Hu 等 (2009) 和 Wang 等 (2011) 利用经验似然方法构造了参数分量的区间估计.

在实际问题分析中, 很多时候可以从实际问题本身或者其他辅助信息那里得到关于模型系数的一些先验信息, 这些信息将有助于提高估计量的有效性, 有关例子可参考 Toutenburg (1982). 对于普通的线性回归模型, 如果对于模型系数有额外的附加条件, 而且附加条件为线性约束条件时, 我们知道有对应的约束最小二乘估计, 该估计在约束条件成立时要比普通的最小二乘估计有效, 具体讨论可参考 Rao 等 (2008). 此外, 对于这些约束条件是否成立我们要进行检验. 就作者所知, 目前关于半参数模型或者变量含误差模型进行约束推断的结果非常少, 其中 Shalabh 等 (2007) 对线性变量含误差模型构造了几类估计, Przystalski 和 Krajewski (2007) 研究了部分线性模型的约束估计问题. Wei,Jia 和 Hu(2013a) 研究了部分线性变量含误差模型的估计和检验问题.

本章将考虑模型 (4.1) 中的参数分量含有约束条件下的估计与检验问题, 本章的结果是对 Shalabh 等 (2007) 和 Przystalski 和 Krajewski (2007) 的推广. 考虑如下的约束条件

$$\boldsymbol{A}\boldsymbol{\beta} = \boldsymbol{b}, \tag{4.2}$$

其中 $\boldsymbol{A}$ 是一个 $k \times p$ 的已知矩阵, $\boldsymbol{b}$ 是一个由 k 个已知数组成的向量, 此外假设 $\text{rank}(\boldsymbol{A}) = k$.

4.2 参数分量的约束估计

假设 $\{Y_i, X_i, Z_i, U_i\}_{i=1}^n$ 是取自模型 (4.1) 的 n 组独立观测数据, 即它们满足

$$Y_i = X_i^{\mathrm{T}}\boldsymbol{\beta} + Z_i^{\mathrm{T}}\boldsymbol{\alpha}(U_i) + \varepsilon_i. \tag{4.3}$$

如果参数分量 $\boldsymbol{\beta}$ 已知, 则模型 (4.3) 转化为如下形式

$$Y_i - X_i^{\mathrm{T}}\boldsymbol{\beta} = \alpha_1(U_i)Z_{i1} + \cdots + \alpha_q(U_i)Z_{iq} + \varepsilon_i. \tag{4.4}$$

显然模型 (4.4) 是一标准的变系数模型. 采用局部线性方法去估计未知的模型系数函数 $\{\alpha_j(\cdot), j = 1, 2, \cdots, q\}$. 给定 u_0 邻域内的一点 u, 对 $\alpha_j(u)$ 利用 Taylor 展开式有

$$\alpha_j(u) \approx \alpha_j(u_0) + \alpha_j'(u_0)(u - u_0), \quad j = 1, 2, \cdots, q, \tag{4.5}$$

其中 $\alpha_j'(u) = \partial\alpha_j(u)/\partial u$. 从而可通过针对 $a_j(u_0), a_j'(u_0)$ 极小化

$$\sum_{i=1}^n \left[\left(Y_i - X_i^{\mathrm{T}}\boldsymbol{\beta}\right) - \sum_{j=1}^q \left\{\alpha_j(u_0) + \alpha_j'(u_0)(U_i - u_0)\right\} Z_{ij} \right]^2 K_h(U_i - u_0) \tag{4.6}$$

来求取它们的估计值. 其中 $K_h(\cdot)=K(\cdot/h)/h$, K 是核函数, h 是窗宽. 经过简单的计算, 可得到

$$\left[\hat{\alpha}_1(u_0),\cdots,\hat{\alpha}_q(u_0),\hat{\alpha}_1'(u_0),\cdots,\hat{\alpha}_q'(u_0)\right]^{\mathrm{T}}=\{\boldsymbol{D}_{u_0}^{\mathrm{T}}\boldsymbol{W}_{u_0}\boldsymbol{D}_{u_0}\}^{-1}\boldsymbol{D}_{u_0}^{\mathrm{T}}\boldsymbol{W}_{u_0}(\boldsymbol{Y}-\boldsymbol{X}\boldsymbol{\beta}),\tag{4.7}$$

其中

$$\boldsymbol{X}=\begin{pmatrix}X_1^{\mathrm{T}}\\X_2^{\mathrm{T}}\\\vdots\\X_n^{\mathrm{T}}\end{pmatrix}=\begin{pmatrix}X_{11}&\cdots&X_{1p}\\X_{21}&\cdots&X_{2p}\\\vdots&&\vdots\\X_{n1}&\cdots&X_{np}\end{pmatrix},\quad \boldsymbol{D}_{u_0}=\begin{pmatrix}Z_1^{\mathrm{T}}&(U_1-u_0)Z_1^{\mathrm{T}}\\Z_2^{\mathrm{T}}&(U_2-u_0)Z_2^{\mathrm{T}}\\\vdots&\vdots\\Z_n^{\mathrm{T}}&(U_n-u_0)Z_n^{\mathrm{T}}\end{pmatrix},$$

以及 $\boldsymbol{Y}=(Y_1,Y_2,\cdots,Y_n)^{\mathrm{T}},\boldsymbol{W}_{u_0}=\mathrm{diag}(K_h(U_1-u_0),K_h(U_2-u_0),\cdots,K_h(U_n-u_0))$.

将模型 (4.3) 中的 $\boldsymbol{\alpha}(U_i)$ 用 $\hat{\boldsymbol{\alpha}}(U_i)$ 代替, 整理可得如下的线性模型

$$Y_i-\hat{Y}_i=(X_i-\hat{X}_i)^{\mathrm{T}}\boldsymbol{\beta}+\varepsilon_i,\tag{4.8}$$

其中 $\hat{\boldsymbol{Y}}=(\hat{Y}_1,\cdots,\hat{Y}_n)^{\mathrm{T}}=\boldsymbol{S}\boldsymbol{Y}$, $\hat{\boldsymbol{X}}=(\hat{X}_1,\cdots,\hat{X}_n)^{\mathrm{T}}=\boldsymbol{S}\boldsymbol{X}$,

$$\boldsymbol{S}=\begin{pmatrix}(Z_1^{\mathrm{T}}\ \boldsymbol{0})(\boldsymbol{D}_{u_1}^{\mathrm{T}}\boldsymbol{W}_{u_1}\boldsymbol{D}_{u_1})^{-1}\boldsymbol{D}_{u_1}^{\mathrm{T}}\boldsymbol{W}_{u_1}\\(Z_2^{\mathrm{T}}\ \boldsymbol{0})(\boldsymbol{D}_{u_2}^{\mathrm{T}}\boldsymbol{W}_{u_2}\boldsymbol{D}_{u_2})^{-1}\boldsymbol{D}_{u_2}^{\mathrm{T}}\boldsymbol{W}_{u_2}\\\vdots\\(Z_n^{\mathrm{T}}\ \boldsymbol{0})(\boldsymbol{D}_{u_n}^{\mathrm{T}}\boldsymbol{W}_{u_n}\boldsymbol{D}_{u_n})^{-1}\boldsymbol{D}_{u_n}^{\mathrm{T}}\boldsymbol{W}_{u_n}\end{pmatrix},$$

基于线性模型 (4.8), 可以得到 $\boldsymbol{\beta}$ 的最小二乘估计. 然而, 由于 X_i 不能精确观测, 因此所得估计量不是可行的. You 和 Chen (2006) 提出了如下的校正 profile 最小二乘估计

$$\hat{\boldsymbol{\beta}}=\arg\min_{\boldsymbol{\beta}\in R^p}\left[(\overline{\boldsymbol{Y}}-\overline{\boldsymbol{V}}\boldsymbol{\beta})^{\mathrm{T}}(\overline{\boldsymbol{Y}}-\overline{\boldsymbol{V}}\boldsymbol{\beta})-n\boldsymbol{\beta}^{\mathrm{T}}\boldsymbol{\Sigma}_\eta\boldsymbol{\beta}\right]=(\overline{\boldsymbol{V}}^{\mathrm{T}}\overline{\boldsymbol{V}}-n\boldsymbol{\Sigma}_\eta)^{-1}\overline{\boldsymbol{V}}^{\mathrm{T}}\overline{\boldsymbol{Y}},\tag{4.9}$$

其中 $\overline{\boldsymbol{Y}}=\boldsymbol{Y}-\hat{\boldsymbol{Y}},\overline{\boldsymbol{V}}=\boldsymbol{V}-\hat{\boldsymbol{V}}$, $\hat{\boldsymbol{V}}=(\hat{V}_1,\cdots,\hat{V}_n)^{\mathrm{T}}=\boldsymbol{S}\boldsymbol{V}$, $\boldsymbol{V}=(V_1,\cdots,V_n)^{\mathrm{T}}$.

下面考虑线性约束 $\boldsymbol{A}\boldsymbol{\beta}=\boldsymbol{b}$ 下模型的估计问题. 基于 Lagrange 乘子法, 构造如下的辅助函数

$$F(\boldsymbol{\beta},\boldsymbol{\lambda})=(\overline{\boldsymbol{Y}}-\overline{\boldsymbol{V}}\boldsymbol{\beta})^{\mathrm{T}}(\overline{\boldsymbol{Y}}-\overline{\boldsymbol{V}}\boldsymbol{\beta})-n\boldsymbol{\beta}^{\mathrm{T}}\boldsymbol{\Sigma}_\eta\boldsymbol{\beta}+2\boldsymbol{\lambda}^{\mathrm{T}}(\boldsymbol{A}\boldsymbol{\beta}-\boldsymbol{b}),\tag{4.10}$$

其中 $\boldsymbol{\lambda}$ 为 k 维 Lagrange 乘子. 针对函数 $\boldsymbol{F}(\boldsymbol{\beta},\boldsymbol{\lambda})$ 关于 $\boldsymbol{\beta},\boldsymbol{\lambda}$ 分别求导, 并令偏导数等于 $\boldsymbol{0}$, 有

$$\frac{\partial F(\boldsymbol{\beta},\boldsymbol{\lambda})}{\partial\boldsymbol{\beta}}=-2\overline{\boldsymbol{V}}^{\mathrm{T}}\overline{\boldsymbol{Y}}-2n\boldsymbol{\Sigma}_\eta\boldsymbol{\beta}+2\overline{\boldsymbol{V}}^{\mathrm{T}}\overline{\boldsymbol{V}}\boldsymbol{\beta}+2\boldsymbol{A}^{\mathrm{T}}\boldsymbol{\lambda}=\boldsymbol{0},\tag{4.11}$$

$$\frac{\partial F(\boldsymbol{\beta},\boldsymbol{\lambda})}{\partial \boldsymbol{\lambda}} = 2(\boldsymbol{A\beta}-\boldsymbol{b}) = \boldsymbol{0}. \tag{4.12}$$

解上述方程, 可得 $\boldsymbol{\beta}$ 的约束 profile 最小二乘估计为

$$\hat{\boldsymbol{\beta}}_r = \hat{\boldsymbol{\beta}} - (\overline{\boldsymbol{V}}^{\mathrm{T}}\overline{\boldsymbol{V}} - n\boldsymbol{\Sigma}_\eta)^{-1}\boldsymbol{A}^{\mathrm{T}}\left[\boldsymbol{A}(\overline{\boldsymbol{V}}^{\mathrm{T}}\overline{\boldsymbol{V}} - n\boldsymbol{\Sigma}_\eta)^{-1}\boldsymbol{A}^{\mathrm{T}}\right]^{-1}(\boldsymbol{A}\hat{\boldsymbol{\beta}}-\boldsymbol{b}). \tag{4.13}$$

此外, $\boldsymbol{M} = \left(Z_1^{\mathrm{T}}\boldsymbol{\alpha}(U_1), Z_2^{\mathrm{T}}\boldsymbol{\alpha}(U_2), \cdots, Z_n^{\mathrm{T}}\boldsymbol{\alpha}(U_n)\right)^{\mathrm{T}}$ 的约束估计为

$$\hat{\boldsymbol{M}}_r = \boldsymbol{S}(\boldsymbol{Y}-\boldsymbol{V}\hat{\boldsymbol{\beta}}_r). \tag{4.14}$$

下面给出 $\hat{\boldsymbol{\beta}}_r$ 的渐近性质. 定义 $\boldsymbol{\Gamma}(u) = E(\boldsymbol{Z}\boldsymbol{Z}^{\mathrm{T}}|U=u)$, $\boldsymbol{\Phi}(u) = E(\boldsymbol{Z}\boldsymbol{X}^{\mathrm{T}}|U=u)$, $A^{\otimes 2}$ 的含义是 AA^{T}.

定理 4.1 假定 4.5 节中的条件 (A.1)~ 条件 (A.5) 成立, $\boldsymbol{\beta}$ 的约束 profile 最小二乘估计是渐近正态的,

$$\sqrt{n}(\hat{\boldsymbol{\beta}}_r - \boldsymbol{\beta}) \xrightarrow{d} N(\boldsymbol{0}, (\boldsymbol{I}_p-\boldsymbol{D})\boldsymbol{\Sigma}_1^{-1}\boldsymbol{\Sigma}_2\boldsymbol{\Sigma}_1^{-1}(\boldsymbol{I}_p-\boldsymbol{D})^{\mathrm{T}}),$$

其中 $\boldsymbol{\Sigma}_1 = E\left\{[X_1 - \boldsymbol{\Phi}^{\mathrm{T}}(U_1)\boldsymbol{\Gamma}^{-1}(U_1)Z_1]^{\otimes 2}\right\}$, $\boldsymbol{D} = \boldsymbol{\Sigma}_1^{-1}\boldsymbol{A}^{\mathrm{T}}[\boldsymbol{A}\boldsymbol{\Sigma}_1^{-1}\boldsymbol{A}^{\mathrm{T}}]^{-1}\boldsymbol{A}$ 和 $\boldsymbol{\Sigma}_2 = E\left\{[X_1 - \boldsymbol{\Phi}^{\mathrm{T}}(U_1)\boldsymbol{\Gamma}^{-1}(U_1)Z_1 + \eta_1](\varepsilon_1 - \eta_1^{\mathrm{T}}\boldsymbol{\beta}) + \boldsymbol{\Sigma}_\eta\boldsymbol{\beta}\right\}^{\otimes 2}$.

为了应用上面的定理进行推断, 要估计 $\hat{\boldsymbol{\beta}}_r$ 的渐近协方差矩阵. 由 You 和 Chen (2006), 可知

$$\hat{\boldsymbol{\Sigma}}_1 = n^{-1}\sum_{i=1}^{n}\left\{(\boldsymbol{V}_i - \hat{\boldsymbol{V}}_i)(\boldsymbol{V}_i - \hat{\boldsymbol{V}}_i)^{\mathrm{T}} - \boldsymbol{\Sigma}_\eta\right\}$$

和

$$\hat{\boldsymbol{\Sigma}}_2 = n^{-1}\sum_{i=1}^{n}\left\{(\boldsymbol{V}_i - \hat{\boldsymbol{V}}_i)[\boldsymbol{Y}_i - \hat{\boldsymbol{Y}}_i - (\boldsymbol{V}_i - \hat{\boldsymbol{V}}_i)^{\mathrm{T}}\hat{\boldsymbol{\beta}}] + \boldsymbol{\Sigma}\boldsymbol{e}\hat{\boldsymbol{\beta}}\right\}^{\otimes 2}$$

分别是 $\boldsymbol{\Sigma}_1$ 和 $\boldsymbol{\Sigma}_2$ 的相合估计. 令 $\hat{\boldsymbol{D}} = \hat{\boldsymbol{\Sigma}}_1^{-1}\boldsymbol{A}^{\mathrm{T}}[\boldsymbol{A}\hat{\boldsymbol{\Sigma}}_1^{-1}\boldsymbol{A}^{\mathrm{T}}]^{-1}\boldsymbol{A}$, 则 $(\boldsymbol{I}_p - \hat{\boldsymbol{D}})\hat{\boldsymbol{\Sigma}}_1^{-1}\hat{\boldsymbol{\Sigma}}_2\hat{\boldsymbol{\Sigma}}_1^{-1}(\boldsymbol{I}_p - \hat{\boldsymbol{D}})^{\mathrm{T}}$ 是 $(\boldsymbol{I}_p - \boldsymbol{D})\boldsymbol{\Sigma}_1^{-1}\boldsymbol{\Sigma}_2\boldsymbol{\Sigma}_1^{-1}(\boldsymbol{I}_p - \boldsymbol{D})^{\mathrm{T}}$ 的相合估计.

4.3 参数分量的检验

对于约束条件, 要检验其适用性. 不失一般性, 考虑如下的假设检验问题

$$H_0: \boldsymbol{A\beta} = \boldsymbol{0} \quad \text{VS} \quad H_0: \boldsymbol{A\beta} \neq \boldsymbol{0}. \tag{4.15}$$

对于部分线性变系数模型, Fan 和 Huang (2005) 提出了一种 profile 广义似然比检验方法去检验参数分量的约束问题. 然而, 由于测量误差的存在, 该方法不能直接推广到模型 (4.1) 上. 我们将构造一种新的检验统计量.

当原假设成立时, 我们能得到 $\boldsymbol{\beta}$ 和 $\boldsymbol{M}$ 的约束估计, 从而定义 H_0 成立时的校正残差平方和为

$$\mathrm{RSS}(H_0) = \| \boldsymbol{Y} - \boldsymbol{V}\hat{\boldsymbol{\beta}}_r - \hat{\boldsymbol{M}}_r \|^2 - n\hat{\boldsymbol{\beta}}_r^{\mathrm{T}} \boldsymbol{\Sigma}_\eta \hat{\boldsymbol{\beta}}_r. \tag{4.16}$$

此外, 有备择假设下的校正残差平方和

$$\mathrm{RSS}(H_1) = \| \boldsymbol{Y} - \boldsymbol{V}\hat{\boldsymbol{\beta}} - \hat{\boldsymbol{M}} \|^2 - n\hat{\boldsymbol{\beta}}^{\mathrm{T}} \boldsymbol{\Sigma}_\eta \hat{\boldsymbol{\beta}}, \tag{4.17}$$

其中 $\hat{\boldsymbol{M}} = \boldsymbol{S}(\boldsymbol{Y} - \boldsymbol{V}\hat{\boldsymbol{\beta}})$. 我们构造如下的检验统计量

$$T_n = \mathrm{RSS}(H_0) - \mathrm{RSS}(H_1). \tag{4.18}$$

显然, 检验统计量 T_n 反映了原假设与备择假设下模型的拟合效果的差异. 若二者有显著差异，则倾向于拒绝原假设 H_0. 因此, 大的 T_n 值趋于拒绝原假设 H_0. 下面给出该检验统计量的渐近分布.

定理 4.2 如果 4.5 节中的条件 (A.1)~ 条件 (A.6) 成立, 那么当原假设 H_0 成立时, 有

$$T_n \xrightarrow{D} l_1\chi_{1,1}^2 + l_2\chi_{2,1}^2 + \cdots + l_k\chi_{k,1}^2,$$

其中 $\chi_{i,1}^2$, $1 \leqslant i \leqslant k$ 是相互独立的服从自由度为 1 的卡方分布的随机变量, 权重 $\{l_i, 1 \leqslant i \leqslant k\}$ 是矩阵 $\left(\boldsymbol{A}\boldsymbol{\Sigma}_1^{-1}\boldsymbol{A}^{\mathrm{T}}\right)^{-1}\left(\boldsymbol{A}\boldsymbol{\Omega}\boldsymbol{A}^{\mathrm{T}}\right)$ 的特征根, $\boldsymbol{\Omega} = \boldsymbol{\Sigma}_1^{-1}\boldsymbol{\Sigma}_2\boldsymbol{\Sigma}_1^{-1}$.

为了利用定理 4.2, 必须首先估计权重 $\{l_i, 1 \leqslant i \leqslant k\}$. 对于服从加权卡方分布的随机变量, Wang 和 Rao (2002) 提出了一种校正方法, 通过该方法可以将服从加权卡方分布的随机变量转化为服从标准卡方分布的随机变量. 基于该校正方法, 将在 T_n 的基础上构造一类校正检验统计量. 定义校正参数为

$$\rho = \left\{\left(\boldsymbol{A}\hat{\boldsymbol{\beta}}\right)^{\mathrm{T}}\left(\boldsymbol{A}\hat{\boldsymbol{\Omega}}\boldsymbol{A}^{\mathrm{T}}\right)^{-1}\boldsymbol{A}\hat{\boldsymbol{\beta}}\right\} \Big/ \left\{\left(\boldsymbol{A}\hat{\boldsymbol{\beta}}\right)^{\mathrm{T}}\left(\boldsymbol{A}\hat{\boldsymbol{\Sigma}}_1\boldsymbol{A}^{\mathrm{T}}\right)^{-1}\boldsymbol{A}\hat{\boldsymbol{\beta}}\right\},$$

其中 $\hat{\boldsymbol{\Omega}} = \hat{\boldsymbol{\Sigma}}_1^{-1}\hat{\boldsymbol{\Sigma}}_2\hat{\boldsymbol{\Sigma}}_1^{-1}$. 定义调整检验统计量为

$$T_n^* = \rho T_n.$$

下面给出该检验统计量的渐近分布.

定理 4.3 如果 4.5 节中的条件 (A.1)~ 条件 (A.6) 成立, 那么当原假设 H_0 成立时, 有

$$T_n^* \xrightarrow{D} \chi_k^2,$$

其中 χ_k^2 为自由度为 k 的卡方分布随机变量.

定理 4.4 如果 4.5 节中的条件 (A.1)~ 条件 (A.6) 成立, 备择假设下有

$$T_n \xrightarrow{D} \chi_k^2(\lambda),$$

其中 $\chi_k^2(\lambda)$ 是自由度为 k 的非中心卡方分布, 非中心参数 $\lambda = \lim\limits_{n\to\infty} n\boldsymbol{\beta}^{\mathrm{T}}\boldsymbol{A}^{\mathrm{T}}\left(\boldsymbol{A\Omega A}^{\mathrm{T}}\right)^{-1}\boldsymbol{A\beta}$.

不难看出, 我们所提出的检验方法同样可以用在其他类型的半参数变量含误差模型上. Wei 和 Wang (2012) 利用该方法重点研究了部分线性可加变量含误差模型.

4.4 数 值 模 拟

在本节中, 将通过数值模拟来验证我们所提方法的有效性. 假定数据来自于如下的部分线性变系数变量含误差模型

$$y_i = x_{1i}\beta_1 + x_{2i}\beta_2 + z_i\alpha(u_i) + \varepsilon_i, v_{1i} = x_{1i} + e_{1i}, v_{2i} = x_{2i} + e_{2i}, \quad i = 1, 2, \cdots, n.$$

其中 $x_{1i} \sim N(0,1)$, $x_{2i} \sim U(-1.5, 1.5)$, $z_i \sim N(1,1)$, $u_i \sim U(0,1)$, $e_{1i} \sim N(0, 0.5^2)$, $e_{2i} \sim N(0, 0.5^2)$, $\alpha(u_i) = \sin(2\pi u_i)$. 为了简便起见, 进一步假设 $\mathrm{Cov}(e_{1i}, e_{2i}) = 0$. 为了考察模型误差的分布对所提方法的影响, 选取如下都满足 $\sigma^2 = 0.25$ 的四种类型的误差分布: ① $\varepsilon_i \sim N(0, 0.5^2)$, ② $\varepsilon_i \sim U(-\sqrt{3}/2, \sqrt{3}/2)$, ③ $\varepsilon_i \sim \dfrac{1}{8}\chi_8^2 - 1$, ④ $\varepsilon_i \sim \sqrt{3}/4t(8)$. 模拟中选取 $K(x) = \dfrac{15}{16}(1-x^2)^2 I_{|x|\leqslant 1}$, 和窗宽 $h = n^{-1/5}$.

首先考察约束估计的有效性. 取 $\beta_1 = 3, \beta_2 = 2$, 并考虑约束条件 $\beta_1 + \beta_2 = 5$. 利用前面的方法, 能得参数分量的约束估计 $\hat{\boldsymbol{\beta}}^r = (\hat{\beta}_1^r, \hat{\beta}_2^r)^{\mathrm{T}}$ 和无约束估计 $\hat{\boldsymbol{\beta}} = (\hat{\beta}_1, \hat{\beta}_2)^{\mathrm{T}}$. 基于前面的设定, 重复 1000 次, 基于 1000 次所得估计量的均值 (mean), 标准差 (SD) 和均方误差 (MSE) 对约束和无约束估计进行比较. 样本量分别取 $n = 100, 150, 200$. 模拟结果见表 4.1~ 表 4.6. 可以看到在各种情况下约束估计要优于无约束估计, 随着样本量的增加, 两类估计的偏和标准差以及均方误差都在变小.

表 4.1 无约束估计 $\hat{\beta}_1$ 和约束估计 $\hat{\beta}_1^r$ 的比较

误差	n	$\hat{\beta}_1$			$\hat{\beta}_1^r$		
		均值	SD	MSE	均值	SD	MSE
$N(0, 0.5^2)$	100	3.069	0.243	0.064	3.015	0.164	0.027
	150	3.052	0.199	0.042	3.015	0.140	0.019
	200	3.030	0.163	0.027	3.003	0.112	0.012
$U\left(-\frac{\sqrt{3}}{2}, \frac{\sqrt{3}}{2}\right)$	100	3.078	0.254	0.070	3.019	0.172	0.029
	150	3.052	0.199	0.042	3.011	0.135	0.018
	200	3.026	0.162	0.026	3.006	0.118	0.014

续表

误差	n	$\hat{\beta}_1$			$\hat{\beta}_1^r$		
		均值	SD	MSE	均值	SD	MSE
$\frac{1}{8}\chi_8^2-1$	100	3.064	0.245	0.064	3.012	0.171	0.029
	150	3.046	0.187	0.037	3.008	0.131	0.017
	200	3.020	0.162	0.026	2.994	0.116	0.013
$\frac{\sqrt{3}}{4}t(8)$	100	3.076	0.245	0.065	3.011	0.163	0.026
	150	3.040	0.191	0.038	3.005	0.133	0.017
	200	3.033	0.170	0.030	3.007	0.121	0.014

表 4.2 无约束估计 $\hat{\beta}_2$ 和约束估计 $\hat{\beta}_2^r$ 的比较

误差	n	$\hat{\beta}_1$			$\hat{\beta}_1^r$		
		均值	SD	MSE	均值	SD	MSE
$N(0,0.5^2)$	100	2.051	0.291	0.087	1.984	0.164	0.027
	150	2.030	0.222	0.050	1.984	0.140	0.019
	200	2.029	0.189	0.036	1.996	0.112	0.012
$U\left(-\frac{\sqrt{3}}{2},\frac{\sqrt{3}}{2}\right)$	100	2.053	0.289	0.086	1.980	0.172	0.029
	150	2.039	0.230	0.054	1.988	0.135	0.018
	200	2.019	0.189	0.036	1.993	0.118	0.014
$\frac{1}{8}\chi_8^2-1$	100	2.050	0.283	0.083	1.987	0.171	0.029
	150	2.038	0.216	0.048	1.991	0.131	0.017
	200	2.037	0.193	0.038	2.005	0.116	0.013
$\frac{\sqrt{3}}{4}t(8)$	100	2.069	0.275	0.080	1.988	0.163	0.026
	150	2.037	0.225	0.052	1.994	0.133	0.017
	200	2.023	0.199	0.040	1.992	0.121	0.014

表 4.3 1000 次重复下检验 $H_0:\beta_1-\beta_2=0$ 的拒绝频率

c 值	误差		分布	
	$N(0,0.5^2)$	$U\left(-\sqrt{3}/2,\sqrt{3}/2\right)$	$\frac{1}{8}\chi_8^2-1$	$\frac{\sqrt{3}}{4}t(8)$
0.0	0.042	0.041	0.051	0.042
−0.2	0.109	0.087	0.089	0.087
−0.4	0.292	0.270	0.267	0.283
−0.6	0.511	0.504	0.502	0.516
−0.8	0.719	0.745	0.728	0.751
−1.0	0.844	0.864	0.875	0.874
0.2	0.127	0.100	0.096	0.102

续表

c 值	误差		分布	
	$N(0,0.5^2)$	$U\left(-\sqrt{3}/2,\sqrt{3}/2\right)$	$\frac{1}{8}\chi_8^2-1$	$\frac{\sqrt{3}}{4}t(8)$
0.4	0.352	0.326	0.330	0.346
0.6	0.652	0.625	0.620	0.649
0.8	0.881	0.881	0.891	0.882
1.0	0.978	0.964	0.971	0.975

下面考虑如下的假设检验问题

$$H_0:\beta_1-\beta_2=0 \quad \text{VS} \quad H_1:\beta_1-\beta_2\neq 0,$$

其中 $\beta_1=1,\beta_2=1-c, c=0,\pm0.2,\pm0.4,\pm0.6,\pm0.8,\pm1.0$. 对于每种设定情况, 统计在样本量为 50 的情况下重复 1000 次下拒绝原假设的频率. 结果见表 4.7~ 表 4.9. 可以看到随着 c 值越来越大, 拒绝的频率越来越接近 1. 当原假设成立时, 拒绝的频率接近显著性水平.

4.5 定理的证明

下面给出定理需要的一些条件, 这些条件 Fan 和 Huang (2005) 和 You 和 Chen (2006) 也采用过.

(A.1) 随机变量 U 具有有界支撑 $\boldsymbol{\Pi}$, 其密度函数 $f(\cdot)$ 在其支撑上满足 Lipschitz 连续, 且不为 0.

(A.2) 对于任一 $U\in\boldsymbol{\Omega}$, 矩阵 $E(\boldsymbol{x}\boldsymbol{x}^{\mathrm{T}}|U)$ 为非奇异, $E(\boldsymbol{x}\boldsymbol{x}^{\mathrm{T}}|U)$, $E(\boldsymbol{x}\boldsymbol{x}^{\mathrm{T}}|U)^{-1}$ 和 $E(\boldsymbol{x}\boldsymbol{z}^{\mathrm{T}}|U)$ 都是 Lipschitz 连续的.

(A.3) 存在 $s>2$ 使得 $E\|\boldsymbol{x}\|^{2s}<\infty$ 和 $E\|\boldsymbol{z}\|^{2s}<\infty$, 对于 $\varepsilon<2-s^{-1}$ 使得 $n^{2\varepsilon-1}h\to\infty$.

(A.4) $\{\alpha_j(\cdot),j=1,\cdots,q\}$ 二阶连续可导.

(A.5) 函数 $K(\cdot)$ 为对称密度函数, 具有紧支撑.

(A.6) $nh^8\to 0$ 和 $nh^2/(\log n)^2\to\infty$.

引理 4.1　在条件 (A.1)~ 条件 (A.5) 成立的条件下,

$$\frac{1}{n}\left(\overline{\boldsymbol{V}}^{\mathrm{T}}\overline{\boldsymbol{V}}-n\boldsymbol{\Sigma}_\eta\right)\xrightarrow{p}\boldsymbol{\Sigma}_1.$$

该引理来自 You 和 Chen (2006).

引理 4.2　在条件 (A.1)~ 条件 (A.5) 成立的条件下 $\boldsymbol{\beta}$ 的约束校正最小二乘估计是渐近正态的,

$$\sqrt{n}(\hat{\boldsymbol{\beta}}-\boldsymbol{\beta})\xrightarrow{d}N(\boldsymbol{0},\boldsymbol{\Omega}).$$

该引理是 You 和 Chen (2006) 的定理 3.1.

引理 4.3 设 $\boldsymbol{\eta} \xrightarrow{D} N(\mathbf{0}, \boldsymbol{I}_k)$, $\boldsymbol{U}$ 是一个 $k \times k$ 非负定矩阵且其特征值为 $l_1, \cdots, l_k$. 则有

$$\boldsymbol{\eta}^{\mathrm{T}} \boldsymbol{U} \boldsymbol{\eta} \xrightarrow{D} l_1 \chi_{1,1}^2 + l_2 \chi_{2,1}^2 + \cdots + l_k \chi_{k,1}^2,$$

这里 $\chi_{i,1}^2$ 如定理中所定义.

证明 由假设可知存在一个正交矩阵 $\boldsymbol{P}$, 使得 $\boldsymbol{U} = \boldsymbol{P}^{\mathrm{T}} \boldsymbol{D} \boldsymbol{P}$, 其中 $\boldsymbol{D} = \operatorname{diag}(l_1, \cdots, l_k)$ 是一个对角元素为 $l_1, \cdots, l_k$ 的对角矩阵. 设 $\overline{\boldsymbol{\eta}} = \boldsymbol{P}\boldsymbol{\eta} = (\overline{\boldsymbol{\eta}}_1, \cdots, \overline{\boldsymbol{\eta}}_k)^{\mathrm{T}}$, 则有 $\overline{\boldsymbol{\eta}} \xrightarrow{D} N(\mathbf{0}, \boldsymbol{I}_k)$.

$$\boldsymbol{\eta}^{\mathrm{T}} \boldsymbol{U} \boldsymbol{\eta} = (\boldsymbol{P}\boldsymbol{\eta})^{\mathrm{T}} \boldsymbol{D} (\boldsymbol{P}\boldsymbol{\eta}) = l_1 \overline{\boldsymbol{\eta}}_1^2 + \cdots + l_k \overline{\boldsymbol{\eta}}_k^2,$$

其中 $\overline{\boldsymbol{\eta}}_1^2, \cdots, \overline{\boldsymbol{\eta}}_k^2$ 是独立同分布于自由度为 1 的卡方分布的随机变量. 证毕.

定理 4.1 的证明 由引理 4.2, 可得到定理 4.1 的结论. 在此省略.

定理 4.2 的证明 为了叙述方便, 定义

$$\boldsymbol{C} = (\overline{\boldsymbol{V}}^{\mathrm{T}} \overline{\boldsymbol{V}} - n\boldsymbol{\Sigma}_\eta)^{-1} \boldsymbol{A}^{\mathrm{T}} \left[\boldsymbol{A} (\overline{\boldsymbol{V}}^{\mathrm{T}} \overline{\boldsymbol{V}} - n\boldsymbol{\Sigma}_\eta)^{-1} \boldsymbol{A}^{\mathrm{T}} \right]^{-1},$$

则由 $\hat{\boldsymbol{\beta}}_r$ 的定义, 有 $\hat{\boldsymbol{\beta}}_r = \hat{\boldsymbol{\beta}} - \boldsymbol{C}(\boldsymbol{A}\hat{\boldsymbol{\beta}} - \boldsymbol{b})$. 由 $\mathrm{RSS}(H_0)$ 的定义, 有

$$\begin{aligned}
\mathrm{RSS(H_0)} &= \left[\boldsymbol{Y} - \boldsymbol{V}\hat{\boldsymbol{\beta}}_r - \hat{\boldsymbol{M}}_r \right]^{\mathrm{T}} \left[\boldsymbol{Y} - \boldsymbol{V}\hat{\boldsymbol{\beta}}_r - \hat{\boldsymbol{M}}_r \right] - n\hat{\boldsymbol{\beta}}_r^{\mathrm{T}} \boldsymbol{\Sigma}_\eta \hat{\boldsymbol{\beta}}_r \\
&= \left[\overline{\boldsymbol{Y}} - \overline{\boldsymbol{V}}\hat{\boldsymbol{\beta}}_r \right]^{\mathrm{T}} \left[\overline{\boldsymbol{Y}} - \overline{\boldsymbol{V}}\hat{\boldsymbol{\beta}}_r \right] - n\hat{\boldsymbol{\beta}}_r^{\mathrm{T}} \boldsymbol{\Sigma}_\eta \hat{\boldsymbol{\beta}}_r \\
&= \left[\overline{\boldsymbol{Y}} - \overline{\boldsymbol{V}}\hat{\boldsymbol{\beta}} + \overline{\boldsymbol{V}}\boldsymbol{C}(\boldsymbol{A}\hat{\boldsymbol{\beta}} - \boldsymbol{b}) \right]^{\mathrm{T}} \left[\overline{\boldsymbol{Y}} - \overline{\boldsymbol{V}}\hat{\boldsymbol{\beta}} + \overline{\boldsymbol{V}}\boldsymbol{C}(\boldsymbol{A}\hat{\boldsymbol{\beta}} - \boldsymbol{b}) \right] \\
&\quad - n \left[\hat{\boldsymbol{\beta}} - \boldsymbol{C}(\boldsymbol{A}\hat{\boldsymbol{\beta}} - \boldsymbol{b}) \right]^{\mathrm{T}} \boldsymbol{\Sigma}_\eta \left[\hat{\boldsymbol{\beta}} - \boldsymbol{C}(\boldsymbol{A}\hat{\boldsymbol{\beta}} - \boldsymbol{b}) \right] \\
&= \mathrm{RSS(H_1)} + (\boldsymbol{A}\hat{\boldsymbol{\beta}} - \boldsymbol{b})^{\mathrm{T}} \left[\boldsymbol{A} \left(\overline{\boldsymbol{V}}^{\mathrm{T}} \overline{\boldsymbol{V}} - \mathrm{n}\boldsymbol{\Sigma}_\eta \right)^{-1} \mathrm{A}^{\mathrm{T}} \right]^{-1} (\boldsymbol{A}\hat{\boldsymbol{\beta}} - \boldsymbol{b}).
\end{aligned} \tag{4.19}$$

从而, 有

$$T_n = \mathrm{RSS}(H_0) - \mathrm{RSS}(H_1) = (\boldsymbol{A}\hat{\boldsymbol{\beta}} - \boldsymbol{b})^{\mathrm{T}} \left[\boldsymbol{A} \left(\overline{\boldsymbol{V}}^{\mathrm{T}} \overline{\boldsymbol{V}} - n\boldsymbol{\Sigma}_\eta \right)^{-1} \boldsymbol{A}^{\mathrm{T}} \right]^{-1} (\boldsymbol{A}\hat{\boldsymbol{\beta}} - \boldsymbol{b}). \tag{4.20}$$

由引理 4.1 和引理 4.2, 可以得到

$$\frac{1}{n} \boldsymbol{A} \left(\overline{\boldsymbol{V}}^{\mathrm{T}} \overline{\boldsymbol{V}} - n\boldsymbol{\Sigma}_\eta \right)^{-1} \boldsymbol{A}^{\mathrm{T}} \xrightarrow{p} \boldsymbol{A}\boldsymbol{\Sigma}_1^{-1} \boldsymbol{A}^{\mathrm{T}}, \tag{4.21}$$

和

$$\sqrt{n} \left[(\boldsymbol{A}\hat{\boldsymbol{\beta}} - \boldsymbol{b}) - (\boldsymbol{A}\boldsymbol{\beta} - \boldsymbol{b}) \right] \xrightarrow{D} N(\mathbf{0}, \boldsymbol{A}\boldsymbol{\Omega}\boldsymbol{A}^{\mathrm{T}}). \tag{4.22}$$

由式 (4.21), 有

$$
\begin{aligned}
T_n &= \left[\sqrt{n}(\boldsymbol{A}\hat{\boldsymbol{\beta}}-\boldsymbol{b})\right]^{\mathrm{T}} \left(\boldsymbol{A}\boldsymbol{\Sigma}_1^{-1}\boldsymbol{A}^{\mathrm{T}}\right)^{-1} \left[\sqrt{n}(\boldsymbol{A}\hat{\boldsymbol{\beta}}-\boldsymbol{b})\right] + o_p(1) \\
&= \left[(\boldsymbol{A}\boldsymbol{\Omega}\boldsymbol{A}^{\mathrm{T}})^{-1/2}\sqrt{n}(\boldsymbol{A}\hat{\boldsymbol{\beta}}-\boldsymbol{b})\right]^{\mathrm{T}} (\boldsymbol{A}\boldsymbol{\Omega}\boldsymbol{A}^{\mathrm{T}})^{1/2} \left(\boldsymbol{A}\boldsymbol{\Sigma}_1^{-1}\boldsymbol{A}^{\mathrm{T}}\right)^{-1} (\boldsymbol{A}\boldsymbol{\Omega}\boldsymbol{A}^{\mathrm{T}})^{1/2} \\
&\quad \times \left[(\boldsymbol{A}\boldsymbol{\Omega}\boldsymbol{A}^{\mathrm{T}})^{-1/2}\sqrt{n}(\boldsymbol{A}\hat{\boldsymbol{\beta}}-\boldsymbol{b})\right] + o_p(1). \qquad (4.23)
\end{aligned}
$$

由式 (4.22), 在原假设成立的情况下, 有

$$
(\boldsymbol{A}\boldsymbol{\Omega}\boldsymbol{A}^{\mathrm{T}})^{-1/2}\sqrt{n}(\boldsymbol{A}\hat{\boldsymbol{\beta}}-\boldsymbol{b}) \xrightarrow{D} N(\boldsymbol{0}, \boldsymbol{I}_k).
$$

同时注意 $(\boldsymbol{A}\boldsymbol{\Omega}\boldsymbol{A}^{\mathrm{T}})^{1/2}\left(\boldsymbol{A}\boldsymbol{\Sigma}_1^{-1}\boldsymbol{A}^{\mathrm{T}}\right)^{-1}(\boldsymbol{A}\boldsymbol{\Omega}\boldsymbol{A}^{\mathrm{T}})^{1/2}$ 和 $\left(\boldsymbol{A}\boldsymbol{\Sigma}_1^{-1}\boldsymbol{A}^{\mathrm{T}}\right)^{-1}\left(\boldsymbol{A}\boldsymbol{\Omega}\boldsymbol{A}^{\mathrm{T}}\right)$ 具有相同的特征根, 则由引理 4.3 可以得到定理 4.2 的结论.

定理 4.3 的证明 由 T_n^* 的定义, 再加上式 (4.23), 有

$$
T_n^* = \sqrt{n}\hat{\boldsymbol{\beta}}^{\mathrm{T}}\boldsymbol{A}^{\mathrm{T}}(\boldsymbol{A}\hat{\boldsymbol{\Omega}}\boldsymbol{A}^{\mathrm{T}})^{-1}\sqrt{n}\boldsymbol{A}\hat{\boldsymbol{\beta}} + o_p(1). \qquad (4.24)
$$

由 You 和 Chen (2006), 我们知道 $\hat{\boldsymbol{\Omega}}$ 是 $\boldsymbol{\Omega}$ 的相和估计. 再由 You 和 Chen (2006) 定理 3.1, 定理 4.3 得证.

定理 4.4 的证明 类似于定理 4.3 的证明, 可以证明定理 4.4.

第 5 章　部分线性变系数测量误差模型的估计

作为部分线性模型与变系数模型的推广, 部分线性变系数模型是一类应用非常广泛的模型. 本章主要研究当非参数部分的协变量测量含误差时模型的估计问题. 基于校正局部线性方法, 我们给出了模型中参数分量以及非参数分量的估计, 并证明了这些估计量的渐近正态性.

5.1 引　　言

数据分析中, 大量问题涉及两组变量间的关系的研究, 回归分析作为解决此类问题的一个有效的统计方法而得到十分广泛的应用. 为了更好地探求因变量与解释变量之间蕴涵的复杂关系, 借助于计算机强大计算能力而发展起来的非参数回归模型在近二十年来得到了人们的广泛关注. 非参数回归模型克服了参数回归方法主观假定函数形式的缺陷, 充分依靠数据本身, “让数据自己说话”. 然而, 非参数回归方法在处理高维数据时会遇到“维数祸根”问题, 并且在分析实际问题时解释意义不强. 在此情况下, 作为非参数模型与参数模型适当折中的半参数建模方法得到了人们的重视. 正如参数模型一样, 半参数模型也有多种形式, 比如, 可加模型、变系数模型、部分线性模型、单指标模型以及它们的混合形式等. 其中一类重要的模型为如下的部分线性变系数模型

$$Y_i = \boldsymbol{X}_i^{\mathrm{T}}\boldsymbol{\alpha}(U_i) + \boldsymbol{Z}_i^{\mathrm{T}}\boldsymbol{\beta} + \varepsilon_i, \quad i = 1, 2, \cdots, n, \tag{5.1}$$

其中 Y_i 为因变量观测值, $\boldsymbol{X}_i = (X_{i1}, X_{i2}, \cdots, X_{iq})^{\mathrm{T}}$, $\boldsymbol{Z}_i = (Z_{i1}, Z_{i2}, \cdots, Z_{ip})^{\mathrm{T}}$ 和 U_i 为对应的自变量观测值. $\boldsymbol{\beta} = (\beta_1, \beta_2, \cdots, \beta_p)^{\mathrm{T}}$ 为 p 维未知待估参数, $\boldsymbol{\alpha}(\cdot) = (\alpha_1(\cdot), \alpha_1(\cdot), \cdots, \alpha_q(\cdot))^{\mathrm{T}}$ 为一列未知函数. 模型误差 ε_i 为相互独立的随机变量, 有 $E(\varepsilon_i|\boldsymbol{X}_i, \boldsymbol{Z}_i, U_i) = 0$ 和 $\mathrm{Var}(\varepsilon_i|\boldsymbol{X}_i, \boldsymbol{Z}_i, U_i) = \sigma^2$.

显然, 模型 (5.1) 包含变系数模型与部分线性模型两种特殊情况. 当 $\boldsymbol{Z}_i = \mathbf{0}$ 时, 模型 (5.1) 即为变系数模型. 变系数模型的建模思想早已存在, 但直到 Hastie 和 Tibshirani (1993) 的发表才引起了人们的重视. 由于其良好的解释能力, 变系数模型已经得到了广泛的研究, 被成功地应用到非线性时间序列建模、函数型数据和纵向数据分析、空间分析以及金融计量分析等相关问题的研究中. 有关内容可参考 Cai 等 (2000a, 2000b), Fan 和 Zhang (1999), Fotheringham 等 (2002), Hoover 等 (1998), Huang 等 (2004) 等文献. 当 $q = 1, \boldsymbol{X}_i = 1$ 时, 模型 (5.1) 即为部分线性模

型. 部分线性模型自 Engle 等 (1986) 研究气温与用电量的关系时提出以后, 受到了统计学家与计量经济学家的广泛关注, 在理论与应用上都得到了深入的研究. 相关内容可参考 Hardle 等 (2000,2004) 以及其中所提及的文献.

作为变系数模型与部分线性模型的推广, 模型 (5.1) 最近得到了人们的关注, 已经有多种方法提出用以估计模型中的参数和非参数部分. Zhang 等 (2002) 基于局部多项式估计方法最早研究了该模型, Zhou 和 You (2004) 利用小波方法估计了模型中的参数部分以及非参数部分. Xia 等 (2004) 基于局部线性方法提出了一种新的有效估计, Ahmad 等 (2005) 研究了模型的一般级数估计, 并利用该模型讨论了中国制造业生产函数的估计问题, Fan 和 Huang (2005) 提出了 profile 最小二乘估计并且基于广义似然比检验方法研究了该模型的检验问题.

实际数据分析中, 某些变量往往不能精确观测. 变量含误差模型在近二十年来得到了人们的广泛关注, 自变量不能精确观测情形下线性模型与非线性模型的研究可参考 Fuller (1987), Cheng 和 Ness (1995) 以及 Carroll 等 (1999). 近年来, 半参数变量含误差模型的研究开始受到人们的重视, 对于模型 (5.1), You 和 Chen (2006) 讨论了当自变量 $\boldsymbol{Z}_i$ 不能精确观测时模型的估计问题. 不同于以上文献, 本书研究当自变量 $\boldsymbol{X}_i$ 不能精确观测时模型 (5.1) 的估计问题. 我们不能得到 $\boldsymbol{X}_i$ 的精确观测值, 得到的是 $\boldsymbol{V}_i$, 二者有如下关系

$$\boldsymbol{V}_i = \boldsymbol{X}_i + \boldsymbol{\xi}_i, \quad i = 1, 2, \cdots, n, \tag{5.2}$$

其中测量误差 $\boldsymbol{\xi}_i$ 为独立同分布的随机变量, 独立于 $(\boldsymbol{X}_i^{\mathrm{T}}, \boldsymbol{Z}_i^{\mathrm{T}}, U_i, \varepsilon_i)^{\mathrm{T}}$, 均值为零, 协方差矩阵为 $\mathrm{Cov}(\boldsymbol{\xi}) = \boldsymbol{\Sigma}_{\boldsymbol{\xi}}$. 为使模型可识别, 我们在本节中假设 $\boldsymbol{\Sigma}_{\boldsymbol{\xi}}$ 是已知的.

本章剩余部分做如下安排, 5.2 节将基于校正局部线性方法与平均的思想给出模型中参数分量与非参数分量的估计; 5.3 节讨论参数分量与非参数分量估计的渐近正态性; 5.4 将给出主要结论的证明.

5.2　校正局部线性估计

注意到模型 (5.1) 和模型 (5.2) 可以看成一类特殊的变系数变量含误差模型. 对于这类模型, You 等 (2006) 提出了系数函数的校正局部多项式估计并研究了估计量的渐近性质. 下面基于校正局部线性方法和平均的思想给出模型中未知参数 $\boldsymbol{\beta}$ 以及系数函数 $\boldsymbol{\alpha}(\cdot)$ 的估计.

当模型 (5.1) 中的协变量 $\boldsymbol{X}_i$ 能精确观测时, 可以利用 Zhang 等 (2002) 中的方法给出参数部分以及非参数部分的估计. 给定 u_0 邻域内的一点 u, 对 $\alpha_j(u)$ 利用 Taylor 展开式有

$$\alpha_j(u) \approx \alpha_j(u_0) + \alpha_j'(u_0)(u - u_0) \equiv a_j + b_j(u - u_0), \quad j = 1, 2, \cdots, q. \tag{5.3}$$

根据加权最小二乘法, u_0 处对应的未知参数 $\boldsymbol{\theta}(u_0)=\left(\boldsymbol{\alpha}(u_0)^{\mathrm{T}},h\boldsymbol{\alpha}'(u_0)^{\mathrm{T}},\boldsymbol{\beta}^{\mathrm{T}}\right)^{\mathrm{T}}$ 可通过使

$$\sum_{i=1}^{n}\left[Y_i-\sum_{j=1}^{q}\{a_j+b_j(U_i-u_0)\}X_{ij}-\boldsymbol{Z}_i^{\mathrm{T}}\boldsymbol{\beta}\right]^2K_h(U_i-u_0) \tag{5.4}$$

达到最小来予以估计. 其中 $K_h(\cdot)=K(\cdot/h)/h$, K 是核函数, h 是窗宽. 基于式 (5.4), 可得到如下的局部估计方程

$$\begin{aligned}0&=\frac{1}{n}\sum_{i=1}^{n}\boldsymbol{\Phi}_i\left\{\boldsymbol{\theta}(u_0);Y_i,\boldsymbol{X}_i,\boldsymbol{Z}_i,U_i\right\}\\&=\frac{1}{n}\sum_{i=1}^{n}K_h(U_i-u_0)(\boldsymbol{\eta}_i\boldsymbol{\eta}_i^{\mathrm{T}}\boldsymbol{\theta}(u_0)-\boldsymbol{\eta}_iY_i),\end{aligned} \tag{5.5}$$

其中 $\boldsymbol{\eta}_i=\left(\boldsymbol{X}_i^{\mathrm{T}},\dfrac{U_i-u_0}{h}\boldsymbol{X}_i^{\mathrm{T}},\boldsymbol{Z}_i^{\mathrm{T}}\right)^{\mathrm{T}}$.

然而变量 $\boldsymbol{X}_i$ 不能精确观测, 基于以下事实

$$E(\boldsymbol{V}_i\boldsymbol{V}_i^{\mathrm{T}}|\boldsymbol{X}_i,\boldsymbol{Z}_i,U_i,Y_i)=E(\boldsymbol{X}_i\boldsymbol{X}_i^{\mathrm{T}})+\boldsymbol{\Sigma}_{\boldsymbol{\xi}},\quad E(\boldsymbol{V}_i\boldsymbol{Z}_i^{\mathrm{T}}|\boldsymbol{X}_i,\boldsymbol{Z}_i,U_i,Y_i)=E(\boldsymbol{X}_i\boldsymbol{Z}_i^{\mathrm{T}}),$$

能得到如下的局部校正估计方程

$$\begin{aligned}0&=\frac{1}{n}\sum_{i=1}^{n}\boldsymbol{\Phi}_i^*\left\{\boldsymbol{\theta}(u_0);Y_i,\boldsymbol{V}_i,\boldsymbol{Z}_i,U_i\right\}\\&=\frac{1}{n}\sum_{i=1}^{n}K_h(U_i-u_0)\left\{(\boldsymbol{\eta}_i^*\boldsymbol{\eta}_i^{*\mathrm{T}}-\boldsymbol{A}_i^{u_0})\boldsymbol{\theta}(u_0)-\boldsymbol{\eta}_iY_i\right\},\end{aligned} \tag{5.6}$$

其中 $\boldsymbol{\eta}_i^*$ 等于 $\boldsymbol{\eta}_i$, 只是将后者中的 $\boldsymbol{X}_i$ 替换为 $\boldsymbol{V}_i$,

$$\boldsymbol{A}_i^{u_0}=\begin{pmatrix}\boldsymbol{A}_i^{u_011}&\boldsymbol{0}_{2q\times p}\\\boldsymbol{0}_{p\times 2q}&\boldsymbol{0}_{p\times p}\end{pmatrix},\quad \boldsymbol{A}_i^{u_011}=\boldsymbol{\Sigma}_{\boldsymbol{\xi}}\otimes\begin{pmatrix}1&(U_i-u)/h\\(U_i-u)/h&[(U_i-u)/h]^2\end{pmatrix}.$$

基于估计方程 (5.6), 得到 $\boldsymbol{\theta}(u_0)$ 的校正局部线性估计为

$$\begin{aligned}\hat{\boldsymbol{\theta}}(u_0)&=\left\{\frac{1}{n}\sum_{i=1}^{n}K_h(U_i-u_0)(\boldsymbol{\eta}_i^*\boldsymbol{\eta}_i^{*\mathrm{T}}-\boldsymbol{A}_i^{u_0})\right\}^{-1}\left\{\frac{1}{n}\sum_{i=1}^{n}K_h(U_i-u_0)\boldsymbol{\eta}_i^*Y_i\right\}\\&=\left\{\boldsymbol{D}_{u_0}^{\mathrm{T}}\boldsymbol{W}_{u_0}\boldsymbol{D}_{u_0}-\boldsymbol{\Omega}_{u_0}\right\}^{-1}\boldsymbol{D}_{u_0}^{\mathrm{T}}\boldsymbol{W}_{u_0}\boldsymbol{Y},\end{aligned} \tag{5.7}$$

其中 $\boldsymbol{W}_{u_0} = \mathrm{diag}(K_h(U_1-u_0), K_h(U_2-u_0), \cdots, K_h(U_n-u_0))$ 以及

$$\boldsymbol{Y} = \begin{pmatrix} Y_1 \\ Y_2 \\ \vdots \\ Y_n \end{pmatrix}, \quad \boldsymbol{D}_{u_0} = \begin{pmatrix} \boldsymbol{V}_1^{\mathrm{T}} & \dfrac{U_1-u_0}{h}\boldsymbol{V}_1^{\mathrm{T}} & \boldsymbol{Z}_1^{\mathrm{T}} \\ \boldsymbol{V}_2^{\mathrm{T}} & \dfrac{U_2-u_0}{h}\boldsymbol{V}_2^{\mathrm{T}} & \boldsymbol{Z}_2^{\mathrm{T}} \\ \vdots & \vdots & \vdots \\ \boldsymbol{V}_n^{\mathrm{T}} & \dfrac{U_n-u_0}{h}\boldsymbol{V}_n^{\mathrm{T}} & \boldsymbol{Z}_n^{\mathrm{T}} \end{pmatrix}, \quad \boldsymbol{\Omega}_{u_0} = \begin{pmatrix} \boldsymbol{\Omega}_{u_0}^{11} & \boldsymbol{0}_{2q\times p} \\ \boldsymbol{0}_{p\times 2q} & \boldsymbol{0}_{p\times p} \end{pmatrix},$$

$$\boldsymbol{\Omega}_{u_0}^{11} = \sum_{i=1}^{n} \boldsymbol{\Sigma}_{\boldsymbol{\xi}} \otimes \begin{pmatrix} 1 & (U_i-u_0)/h \\ (U_i-u_0)/h & [(U_i-u_0)/h]^2 \end{pmatrix} K_h(U_i-u_0).$$

用 U_i 代替前面的 u_0, 即得到 U_i 处对应的未知参数 $\boldsymbol{\beta}$ 的估计为

$$\hat{\boldsymbol{\beta}}(U_i) = (\boldsymbol{0}_{2q}, \boldsymbol{I}_p)\{\boldsymbol{D}_{u_i}^{\mathrm{T}}\boldsymbol{W}_{u_i}\boldsymbol{D}_{u_i} - \boldsymbol{\Omega}_{u_i}\}^{-1}\boldsymbol{D}_{u_i}^{\mathrm{T}}\boldsymbol{W}_{u_i}\boldsymbol{Y}, \tag{5.8}$$

基于平均的思想, 定义 $\boldsymbol{\beta}$ 的最终估计为

$$\hat{\boldsymbol{\beta}} = \frac{1}{n}\sum_{i=1}^{n}\hat{\boldsymbol{\beta}}(U_i). \tag{5.9}$$

至于系数函数 $\boldsymbol{\alpha}(\cdot)$ 的估计, 一般可以采用两步方法. 首先基于前面的方法得到 $\boldsymbol{\beta}$ 的估计 $\hat{\beta}$. 从而模型 (5.1) 和模型 (5.2) 可转化为如下标准的变系数变量含误差模型

$$\begin{cases} Y_i^* = \boldsymbol{X}_i^{\mathrm{T}}\boldsymbol{\alpha}(U_i) + \varepsilon_i, & i = 1, 2, \cdots, n, \\ \boldsymbol{V}_i = \boldsymbol{X}_i + \boldsymbol{\xi}_i, & \mathrm{Cov}(\boldsymbol{\xi}_i) = \boldsymbol{\Sigma}_{\boldsymbol{\xi}}. \end{cases} \tag{5.10}$$

其中 $Y_i^* = Y_i - \boldsymbol{Z}_i^{\mathrm{T}}\hat{\boldsymbol{\beta}}$. 对于该模型, 基于 You 等 (2006) 的校正局部线性方法, 可得到 $\boldsymbol{\alpha}(u)$ 的估计为

$$\hat{\boldsymbol{\alpha}}(u) = (\boldsymbol{I}_q \ \ \boldsymbol{0}_q)\{\boldsymbol{B}_u^{\mathrm{T}}\boldsymbol{W}_u\boldsymbol{B}_u - \boldsymbol{\Omega}_u^{11}\}^{-1}\boldsymbol{B}_u^{\mathrm{T}}\boldsymbol{W}_u(\boldsymbol{Y} - \boldsymbol{Z}\hat{\boldsymbol{\beta}}), \tag{5.11}$$

其中

$$\boldsymbol{B}_u = \begin{pmatrix} \boldsymbol{V}_1^{\mathrm{T}} & \dfrac{U_1-u}{h}\boldsymbol{V}_1^{\mathrm{T}} \\ \boldsymbol{V}_2^{\mathrm{T}} & \dfrac{U_2-u}{h}\boldsymbol{V}_2^{\mathrm{T}} \\ \vdots & \vdots \\ \boldsymbol{V}_n^{\mathrm{T}} & \dfrac{U_n-u}{h}\boldsymbol{V}_n^{\mathrm{T}} \end{pmatrix}.$$

5.3 估计的渐近性质

本节将给出参数分量的估计 $\hat{\boldsymbol{\beta}}$ 与非参数部分系数函数的估计 $\hat{\boldsymbol{\alpha}}(u)$ 的渐近性质. 在给出具体的结果之前, 下面给出一些假设条件. 这些假设只是为了证明的方便, 并非最弱条件. 另外记 $\boldsymbol{\Upsilon}(u) = E(\boldsymbol{Z}\boldsymbol{Z}^{\mathrm{T}}|U = u)$, $\boldsymbol{\Gamma}(u) = E(\boldsymbol{X}\boldsymbol{X}^{\mathrm{T}}|U = u)$, $\Phi(u) = E(\boldsymbol{X}\boldsymbol{Z}^{\mathrm{T}}|U = u), \mu_i = \int_0^{\infty} t^i K(t)\mathrm{d}t, \nu_i = \int_0^{\infty} t^i K^2(t)\mathrm{d}t$, $A^{\otimes 2}$ 表示 AA^{T}.

(A.1) 随机变量 U 具有有界支撑 $\prod$, 其密度函数 $f(\cdot)$ 在其支撑上满足 Lipschitz 连续, 且不为 0.

(A.2) 对于任一 $U \in \boldsymbol{\Omega}$, 矩阵 $E(\boldsymbol{x}\boldsymbol{x}^{\mathrm{T}}|U)$ 为非奇异, $E(\boldsymbol{x}\boldsymbol{x}^{\mathrm{T}}|U)$, $E(\boldsymbol{x}\boldsymbol{x}^{\mathrm{T}}|U)^{-1}$ 和 $E(\boldsymbol{x}\boldsymbol{z}^{\mathrm{T}}|U)$ 都是 Lipschitz 连续的.

(A.3) 存在 $s > 2$ 使得 $E\|\boldsymbol{x}\|^{2s} < \infty$ 和 $E\|\boldsymbol{z}\|^{2s} < \infty$, 对于 $\varepsilon < 2 - s^{-1}$, 使得 $n^{2\varepsilon-1}h \to \infty$.

(A.4) $\{\alpha_j(\cdot), j = 1, \cdots, q\}$ 二阶连续可导.

(A.5) 函数 $K(\cdot)$ 为对称密度函数, 具有紧支撑.

(A.6) $nh^8 \to 0$ 和 $nh^2/(\log n)^2 \to \infty$.

下面给出常值系数 $\boldsymbol{\beta}$ 的估计 $\hat{\boldsymbol{\beta}}$ 的渐近性质.

定理 5.1 若假定 (A.1)∼ 假定 (A.6) 成立, 如果 $h \propto n^r, -1 < r < -1/4$, 那么 $\boldsymbol{\beta}$ 的估计 $\hat{\boldsymbol{\beta}}$ 是渐近正态的, 即

$$\sqrt{n}(\hat{\boldsymbol{\beta}} - \boldsymbol{\beta}) \xrightarrow{d} N(\boldsymbol{0}, \boldsymbol{\Omega}),$$

其中 $\boldsymbol{\Sigma} = \boldsymbol{\Upsilon}(U_1) - \boldsymbol{\Phi}^{\mathrm{T}}(U_1)\boldsymbol{\Gamma}^{-1}(U_1)\boldsymbol{\Phi}(U_1)$, 和

$$\begin{aligned}\boldsymbol{\Omega} =& E\left(\boldsymbol{\Sigma}^{-1}\left\{[\boldsymbol{Z}_1 - \boldsymbol{\Phi}^{\mathrm{T}}(U_1)\boldsymbol{\Gamma}^{-1}(U_1)\boldsymbol{V}_1][\varepsilon_1 - \boldsymbol{\xi}_1^{\mathrm{T}}\boldsymbol{\alpha}(U_1)]\right.\right.\\ &\left.\left.- \boldsymbol{\Phi}^{\mathrm{T}}(U_1)\boldsymbol{\Gamma}^{-1}(U_1)\boldsymbol{\Sigma}_{\boldsymbol{\xi}}\boldsymbol{\alpha}(U_1)\right\}^{\otimes 2}\boldsymbol{\Sigma}^{-1}\right).\end{aligned}$$

注 5.1 该结论表明如果选择一般的窗宽 $h \sim n^{-1/5}$, 那么常值系数 $\boldsymbol{\beta}$ 的估计不是 $\sqrt{n}$ 相合的.

下面给出系数函数的估计 $\hat{\boldsymbol{\alpha}}(u_0)$ 的渐近性质.

定理 5.2 若假定 (A.1)∼ 假定 (A.6) 成立, 那么有

$$\sqrt{nh}\left\{\hat{\boldsymbol{\alpha}}(u) - \boldsymbol{\alpha}(u) - \frac{h^2}{2}\mu_2(K)\boldsymbol{\alpha}''(u) + o_p(h^2)\right\} \xrightarrow{d} N\left(\boldsymbol{0}, \frac{\nu_0(K)}{f(u)}\boldsymbol{\Gamma}(u)^{-1}\boldsymbol{\Omega}(u)\boldsymbol{\Gamma}(u)^{-1}\right),$$

其中 $\boldsymbol{\zeta}_1 = \boldsymbol{\Sigma}_{\boldsymbol{\xi}} - \boldsymbol{\xi}_1\boldsymbol{\xi}_1^{\mathrm{T}} - \boldsymbol{X}_1\boldsymbol{\xi}_1^{\mathrm{T}}$, $\boldsymbol{\Omega}(u) = \sigma^2\boldsymbol{\Gamma}(u) + \sigma^2\boldsymbol{\Sigma}_{\boldsymbol{\xi}} + E[\boldsymbol{\zeta}_1\boldsymbol{\alpha}(u)\boldsymbol{\alpha}^{\mathrm{T}}(u)\boldsymbol{\zeta}_1^{\mathrm{T}}|U_1 = u]$.

注 5.2 由上面的结论可知, 系数函数估计的渐近性质与变系数变量含误差模型中 (即模型 (5.1) 和模型 (5.2) 中 $\boldsymbol{Z}_i = 0$) 系数函数估计的渐近性质是相同的. 这是由于未知参数 $\boldsymbol{\beta}$ 的估计是 $\sqrt{n}$ 相合的, 因此线性部分的存在对非参数部分估计渐近性质的影响是可以忽略的. 这也与部分线性模型, 部分可加模型等半参数模型的有关结论相一致.

5.4 定理的证明

本节将给出有关定理的证明. 在给出具体证明之前, 首先给出如下引理. 记 $c_n = h^2 + \left\{\dfrac{\log(1/h)}{nh}\right\}^{1/2}$.

引理 5.1 令 $(\boldsymbol{X}_1, \boldsymbol{Y}_1), \cdots, (\boldsymbol{X}_n, \boldsymbol{Y}_n)$ 为独立同分布 (iid) 的随机序列, 其中 $\boldsymbol{Y}_i, i = 1, 2, \cdots, n$ 为一元随机变量, 进一步假定 $E|y|^s < \infty$ 与 $\sup_x \displaystyle\int |y|^s f(x, y)\mathrm{d}y < \infty$, 其中 f 表示 $(\boldsymbol{X}, \boldsymbol{Y})$ 的联合密度. 令 K 为一有界正函数, 并有有界支撑且满足 Lipschitz 条件, 则有

$$\sup_x \left|\frac{1}{n}\sum_{i=1}^{n}[K_h(\boldsymbol{X}_i - x)\boldsymbol{Y}_i - E(K_h(\boldsymbol{X}_i - x)\boldsymbol{Y}_i)]\right| = O_p\left(\left\{\frac{\log(1/h)}{nh}\right\}^{1/2}\right),$$

其中对于 $\varepsilon < 1 - s^{-1}$ 有 $n^{2\varepsilon-1}h \to \infty$.

该引理由 Mack 和 Silverman (1999) 即可证得.

引理 5.2 若假定 (A.1)~ 假定 (A.6) 成立, 有

$$\sup_{u\in\boldsymbol{\mathit{\Pi}}} \frac{1}{nh}\sum_{i=1}^{n} K\left(\frac{U_i - u}{h}\right)\left(\frac{U_i - u}{h}\right)^k R_{ij}\varepsilon_i = O_p\left\{\left(\frac{\log n}{nh}\right)^{1/2}\right\},$$

和

$$\sup_{u\in\boldsymbol{\mathit{\Pi}}} \frac{1}{nh}\sum_{i=1}^{n} K\left(\frac{U_i - u}{h}\right)\left(\frac{U_i - u}{h}\right)^k R_{ij}\xi_{ij} = O_p\left\{\left(\frac{\log n}{nh}\right)^{1/2}\right\}.$$

其中 $\boldsymbol{R} = \boldsymbol{X}, \boldsymbol{Z}$.

该引理即为 You 等 (2006) 中的引理 8.1.

定理 5.1 的证明 由 $\alpha_j(\cdot)$ 有连续的二阶导数, 基于 Taylor 展开式, 有

$$\begin{aligned} Y_i &= \boldsymbol{X}_i^{\mathrm{T}}\boldsymbol{\alpha}(U_i) + \boldsymbol{Z}_i^{\mathrm{T}}\boldsymbol{\beta} + \varepsilon_i \\ &= \boldsymbol{X}_i^{\mathrm{T}}\boldsymbol{\alpha}(u) + \boldsymbol{X}_i^{\mathrm{T}}h\boldsymbol{\alpha}'(u)\frac{U_i - u}{h} + \frac{h^2}{2}\boldsymbol{X}_i^{\mathrm{T}}\boldsymbol{\alpha}''(u)(\frac{U_i - u}{h})^2 + o_p(h^2) + \boldsymbol{Z}_i^{\mathrm{T}}\boldsymbol{\beta} + \varepsilon_i. \end{aligned}$$

从而有

$$\boldsymbol{D}_u^{\mathrm{T}}\boldsymbol{W}_u\boldsymbol{Y} = \{\boldsymbol{D}_u^{\mathrm{T}}\boldsymbol{W}_u\boldsymbol{D}_u - \boldsymbol{\varOmega}_u\}\boldsymbol{\theta}(u)$$

$$
\begin{aligned}
&+ \boldsymbol{D}_u^{\mathrm{T}} \boldsymbol{W}_u [\boldsymbol{\varepsilon} - \boldsymbol{B}\boldsymbol{\theta}(u)] + \boldsymbol{\Omega}_u \boldsymbol{\theta}(u) \\
&+ \boldsymbol{D}_u^{\mathrm{T}} \boldsymbol{W}_u \boldsymbol{A} \frac{h^2 \boldsymbol{\alpha}''(u)}{2} + \boldsymbol{D}_u^{\mathrm{T}} \boldsymbol{W}_u \mathbf{1}_n o_p(h^2).
\end{aligned}
$$

由 $\hat{\boldsymbol{\beta}}(u)$ 的表达式, 能得到

$$
\begin{aligned}
\hat{\boldsymbol{\beta}}(u) - \boldsymbol{\beta} &= (\mathbf{0}_{2q},\ \boldsymbol{I}_p) \{\boldsymbol{D}_u^{\mathrm{T}} \boldsymbol{W}_u \boldsymbol{D}_u - \boldsymbol{\Omega}_u\}^{-1} \boldsymbol{D}_u^{\mathrm{T}} \boldsymbol{W}_u \boldsymbol{Y} - \boldsymbol{\beta} \\
&= (\mathbf{0}_{2q},\ \boldsymbol{I}_p) \{\boldsymbol{D}_u^{\mathrm{T}} \boldsymbol{W}_u \boldsymbol{D}_u - \boldsymbol{\Omega}_u\}^{-1} \boldsymbol{D}_u^{\mathrm{T}} \boldsymbol{W}_u \left(\boldsymbol{A}_u \frac{h^2 \boldsymbol{\alpha}''(u)}{2} + \mathbf{1}_n o_p(h^2) \right) \\
&\quad + (\mathbf{0}_{2q},\ \boldsymbol{I}_p) \{\boldsymbol{D}_u^{\mathrm{T}} \boldsymbol{W}_u \boldsymbol{D}_u - \boldsymbol{\Omega}_u\}^{-1} \left\{ \boldsymbol{D}_u^{\mathrm{T}} \boldsymbol{W}_u [\boldsymbol{\varepsilon} - \boldsymbol{B}_u \boldsymbol{\theta}(u)] + \boldsymbol{\Omega}_u \boldsymbol{\theta}(u) \right\},
\end{aligned}
$$

其中

$$
\boldsymbol{A}_u = \begin{pmatrix} \boldsymbol{X}_1^{\mathrm{T}} (\frac{U_1 - u}{h})^2 \\ \boldsymbol{X}_2^{\mathrm{T}} (\frac{U_2 - u}{h})^2 \\ \vdots \\ \boldsymbol{X}_n^{\mathrm{T}} (\frac{U_n - u}{h})^2 \end{pmatrix}, \quad \boldsymbol{B}_u = \begin{pmatrix} \boldsymbol{\xi}_1^{\mathrm{T}} & \frac{U_1 - u}{h} \boldsymbol{\xi}_1^{\mathrm{T}} & \mathbf{0} \\ \boldsymbol{\xi}_2^{\mathrm{T}} & \frac{U_2 - u}{h} \boldsymbol{\xi}_2^{\mathrm{T}} & \mathbf{0} \\ \vdots & \vdots & \vdots \\ \boldsymbol{\xi}_n^{\mathrm{T}} & \frac{U_n - u}{h} \boldsymbol{\xi}_n^{\mathrm{T}} & \mathbf{0} \end{pmatrix}.
$$

从而

$$
\begin{aligned}
&\hat{\boldsymbol{\beta}} - \boldsymbol{\beta} \\
&= \frac{1}{n} \sum_{i=1}^{n} \left(\hat{\boldsymbol{\beta}}(U_i) - \boldsymbol{\beta} \right) \\
&= \frac{1}{n} \sum_{i=1}^{n} \left\{ (\mathbf{0}_{2q},\ \boldsymbol{I}_p) \{\boldsymbol{D}_{u_i}^{\mathrm{T}} \boldsymbol{W}_{u_i} \boldsymbol{D}_{u_i} - \boldsymbol{\Omega}_{u_i}\}^{-1} \boldsymbol{D}_{u_i}^{\mathrm{T}} \boldsymbol{W}_{u_i} \boldsymbol{Y} - \boldsymbol{\beta} \right\} \\
&= \frac{1}{n} \sum_{i=1}^{n} \left\{ (\mathbf{0}_{2q},\ \boldsymbol{I}_p) \{\boldsymbol{D}_{u_i}^{\mathrm{T}} \boldsymbol{W}_{u_i} \boldsymbol{D}_{u_i} - \boldsymbol{\Omega}_{u_i}\}^{-1} \boldsymbol{D}_{u_i}^{\mathrm{T}} \boldsymbol{W}_{u_i} \left[\boldsymbol{A}_{u_i} \frac{h^2 \boldsymbol{\alpha}''(U_i)}{2} + \mathbf{1}_n o_p(h^2) \right] \right\} \\
&\quad + \frac{1}{n} \sum_{i=1}^{n} \left\{ (\mathbf{0}_{2q},\ \boldsymbol{I}_p) \{\boldsymbol{D}_{u_i}^{\mathrm{T}} \boldsymbol{W}_{u_i} \boldsymbol{D}_{u_i} - \boldsymbol{\Omega}_{u_i}\}^{-1} \left[\boldsymbol{D}_{u_i}^{\mathrm{T}} \boldsymbol{W}_{u_i} (\boldsymbol{\varepsilon} - \boldsymbol{B}\boldsymbol{\theta}(u_i)) + \boldsymbol{\Omega}_{u_i} \boldsymbol{\theta}(u_i) \right] \right\} \\
&\doteq I_1 + I_2.
\end{aligned}
\tag{5.12}
$$

基于引理 5.1, 经过计算有如下的结论

$$
\frac{\boldsymbol{D}_u^{\mathrm{T}} \boldsymbol{W}_u \boldsymbol{D}_u - \boldsymbol{\Omega}_u}{n} = f(u) \begin{pmatrix} \boldsymbol{\Gamma}(u) \otimes \begin{pmatrix} 1 & 0 \\ 0 & \mu_2 \end{pmatrix} & \boldsymbol{\Phi}(u) \otimes \begin{pmatrix} 1 \\ 0 \end{pmatrix} \\ \boldsymbol{\Phi}^{\mathrm{T}}(u) \otimes (1, 0) & \boldsymbol{\Upsilon}(u) \end{pmatrix} + o_p(1),
$$

和

$$
\frac{1}{n} \boldsymbol{D}_u^{\mathrm{T}} \boldsymbol{W}_u \boldsymbol{A}_u = f(u) \begin{pmatrix} \boldsymbol{\Gamma}(u) \otimes \begin{pmatrix} \mu_2 \\ 0 \end{pmatrix} \\ \boldsymbol{\Phi}^{\mathrm{T}}(u) \mu_2 \end{pmatrix} + o_p(1).
$$

基于上面的结论, 经过简单的矩阵运算, 有

$$I_1 = o_p(h^2).$$

下面讨论 I_2, 首先记

$$\boldsymbol{D}_{u_i}^{\mathrm{T}} \boldsymbol{W}_{u_i} (\boldsymbol{\varepsilon} - \boldsymbol{B\theta}(u_i)) + \boldsymbol{\Omega}_{u_i} \boldsymbol{\theta}(u_i) = \boldsymbol{C}_{i1} + \boldsymbol{C}_{i2},$$

其中

$$\boldsymbol{C}_{i1} = \begin{pmatrix} \displaystyle\sum_{j=1}^{n} \left\{ \boldsymbol{V}_j[\varepsilon_j - \boldsymbol{\xi}_j^{\mathrm{T}} \boldsymbol{\alpha}(U_i)] + \boldsymbol{\Sigma}_{\boldsymbol{\xi}} \boldsymbol{\alpha}(U_i) \right\} K_h(U_j - U_i) \\ \displaystyle\sum_{j=1}^{n} \left\{ \boldsymbol{V}_j[\varepsilon_j - \boldsymbol{\xi}_j^{\mathrm{T}} \boldsymbol{\alpha}(U_i)] + \boldsymbol{\Sigma}_{\boldsymbol{\xi}} \boldsymbol{\alpha}(U_i) \right\} \frac{U_j - U_i}{h} K_h(U_j - U_i) \\ \displaystyle\sum_{i=1}^{n} \boldsymbol{Z}_j[\varepsilon_j - \boldsymbol{\xi}_j^{\mathrm{T}} \boldsymbol{\alpha}(U_i)] K_h(U_j - U_i) \end{pmatrix}$$

与

$$\boldsymbol{C}_{i2} = \begin{pmatrix} \displaystyle\sum_{j=1}^{n} \left(\boldsymbol{\Sigma}_{\boldsymbol{\xi}} - \boldsymbol{V}_j \boldsymbol{\xi}_j^{\mathrm{T}} \right) \frac{U_j - U_i}{h} K_h(U_j - U_i) h\alpha'(U_i) \\ \displaystyle\sum_{j=1}^{n} \left(\boldsymbol{\Sigma}_{\boldsymbol{\xi}} - \boldsymbol{V}_j \boldsymbol{\xi}_j^{\mathrm{T}} \right) \left(\frac{U_j - U_i}{h} \right)^2 K_h(U_j - U_i) h\alpha'(U_i) \\ \displaystyle -\sum_{j=1}^{n} \boldsymbol{Z}_j \boldsymbol{\xi}_j^{\mathrm{T}} \frac{U_j - U_i}{h} K_h(U_j - U_i) h\alpha'(U_i) \end{pmatrix},$$

由引理 5.1 和 5.2, 有

$$\begin{aligned} &\frac{1}{n} \sum_{i=1}^{n} (\boldsymbol{0}_{2q},\ \boldsymbol{I}_p) \{ \boldsymbol{D}_{u_i}^{\mathrm{T}} \boldsymbol{W}_{u_i} \boldsymbol{D}_{u_i} - \boldsymbol{\Omega}_{u_i} \}^{-1} \boldsymbol{C}_{i1} \\ =& \frac{1}{n} \sum_{i=1}^{n} f(u_i)^{-1} \boldsymbol{\Sigma}_i^{-1} \Bigg\{ \sum_{j=1}^{n} \Bigg\{ [\boldsymbol{Z}_j - \boldsymbol{\Phi}^{\mathrm{T}}(U_i) \boldsymbol{\Gamma}^{-1}(U_i) \boldsymbol{V}_j][\varepsilon_j - \boldsymbol{\xi}_j^{\mathrm{T}} \boldsymbol{\alpha}(U_i)] \\ & - \boldsymbol{\Phi}^{\mathrm{T}}(U_i) \boldsymbol{\Gamma}^{-1}(U_i) \boldsymbol{\Sigma}_{\boldsymbol{\xi}} \boldsymbol{\alpha}(U_i) \Bigg\} K_h(U_j - U_i) \Bigg\} \\ =& \frac{1}{n} \sum_{j=1}^{n} \boldsymbol{\Sigma}_j^{-1} \Bigg\{ [\boldsymbol{Z}_j - \boldsymbol{\Phi}^{\mathrm{T}}(U_j) \boldsymbol{\Gamma}^{-1}(U_j) \boldsymbol{V}_j][\varepsilon_j - \boldsymbol{\xi}_j^{\mathrm{T}} \boldsymbol{\alpha}(U_j)] \\ & - \boldsymbol{\Phi}^{\mathrm{T}}(U_j) \boldsymbol{\Gamma}^{-1}(U_j) \boldsymbol{\Sigma}_{\boldsymbol{\xi}} \boldsymbol{\alpha}(U_j) \Bigg\} + o_p(n^{-1/2}), \end{aligned}$$

以及

$$\frac{1}{n} \sum_{i=1}^{n} (\boldsymbol{0}_{2q},\ \boldsymbol{I}_p) \{ \boldsymbol{D}_{u_i}^{\mathrm{T}} \boldsymbol{W}_{u_i} \boldsymbol{D}_{u_i} - \boldsymbol{\Omega}_{u_i} \}^{-1} \boldsymbol{C}_{i1} = O_p(h^2) O_p \left\{ \left(\frac{\log n}{nh} \right)^{1/2} \right\}.$$

基于以上结果, 有

$$\sqrt{n}(\hat{\boldsymbol{\beta}}-\boldsymbol{\beta}) = \frac{1}{\sqrt{n}}\sum_{j=1}^{n}\boldsymbol{\Sigma}_j^{-1}\Big\{[\boldsymbol{Z}_j-\boldsymbol{\Phi}^{\mathrm{T}}(U_j)\boldsymbol{\Gamma}^{-1}(U_j)\boldsymbol{V}_j][\varepsilon_j-\boldsymbol{\xi}_j^{\mathrm{T}}\boldsymbol{\alpha}(U_j)] - \boldsymbol{\Phi}^{\mathrm{T}}(U_j)\boldsymbol{\Gamma}^{-1}(U_j)\boldsymbol{\Sigma}_{\boldsymbol{\xi}}\boldsymbol{\alpha}(U_j)\Big\}+o_p(1).$$

则由中心极限定理, 可得

$$\sqrt{n}(\hat{\boldsymbol{\beta}}-\boldsymbol{\beta}) \xrightarrow{d} N(\boldsymbol{0},\boldsymbol{\Omega}).$$

定理 5.2 的证明 类似于 You 等 (2006) 中定理 2.1 的证明, 再加上本书定理 5.1 的结论, 可以证明定理 5.2 的成立, 证明细节在此省略.

第 6 章　因变量缺失下部分线性变系数变量含误差模型的估计

近年来, 一方面半参数变量含误差模型的研究已经得到了关注; 另一方面因变量缺失下的半参数模型的研究也已经有了较多的研究成果. 然而, 同时考虑自变量不能精确观测和因变量缺失两种情形的研究还相对较少. 本章主要考虑部分线性变系数模型在自变量含有测量误差以及因变量存在缺失情形下的估计问题, 主要内容来自于文献魏传华 (2010).

6.1　引　　言

近二十年来, 借助于计算机强大计算能力而发展起来的半参数建模方法得到了人们的广泛关注. 半参数模型具有多种形式, 常见的有可加模型、变系数模型、部分线性模型、单指标模型以及它们的混合形式等, 其中一类重要的半参数模型为如下的部分线性变系数模型

$$Y = \boldsymbol{Z}^{\mathrm{T}}\boldsymbol{\beta} + \boldsymbol{X}^{\mathrm{T}}\boldsymbol{\alpha}(U) + \varepsilon, \tag{6.1}$$

其中 Y 是因变量, $\boldsymbol{Z}, \boldsymbol{X}$ 和 U 为自变量, 假设 U 为一单变量. $\boldsymbol{\beta} = (\beta_1, \beta_2, \cdots, \beta_q)^{\mathrm{T}}$ 为未知的参数分量 $\boldsymbol{\alpha}(\cdot) = (\alpha_1(\cdot), \alpha_1(\cdot), \cdots, \alpha_p(\cdot))^{\mathrm{T}}$ 为未知的系数函数, 模型误差 ε 有 $E[\varepsilon|\boldsymbol{X}, \boldsymbol{Z}, U] = 0$.

显然, 当 $\boldsymbol{Z} = 0$ 时, 模型 (6.1) 即为变系数模型. 变系数模型的建模思想早已存在, 但直到 Hastie 和 Tibshirani (1993) 的发表才引起了人们的重视. 由于其良好的解释能力, 变系数模型已经得到了广泛的研究, 被成功地应用到非线性时间序列建模、函数型数据和纵向数据分析, 空间分析以及金融计量分析等相关问题的研究中. 当 $p = 1, \boldsymbol{X} = 1$ 时, 模型 (6.1) 即为部分线性模型. 部分线性模型自 Engle 等 (1986) 研究气温与用电量的关系时提出以后, 受到了统计学家与计量经济学家的广泛关注, 在理论与应用上都得到了深入的研究.

作为变系数模型与部分线性模型的推广, 模型 (6.1) 在近年来得到了人们的重视. Zhang 等 (2002) 基于局部多项式光滑技术提出了模型的两步估计方法, Zhou 和 You (2004) 构造了参数分量以及非参数分量的小波估计, Xia 等 (2004) 基于局部线性方法提出了一种新的有效估计, Ahmad 等 (2005) 给出了模型的一般级数估

计, Fan 和 Huang (2005) 针对参数分量提出了 profile 最小二乘估计并且基于广义似然比检验方法研究了该模型的检验问题. You 和 Chen (2006) 研究了当自变量 **Z** 测量含误差时模型的估计.

实际问题分析中, 会经常遇到数据缺失现象, 缺失数据的处理一直受到统计学家的重视, Little 和 Rubin (2002) 对缺失数据问题进行了深入全面的讨论. 在回归分析领域, Abu-Salin (1988), Cheng (1990), Chu 和 Cheng (1995), Wang 等 (2004), Wang 和 Sun (2007) 以及 Wei (2012b) 等文献讨论了多类回归模型在因变量存在缺失情形时的统计推断问题. 值得注意的是, 这些研究一般都假设回归模型中的自变量能精确观测. 然而, 很多情况下得不到自变量的精确观测值, 只能得到含有误差的观测值. 因此, 变量含误差模型在因变量缺失下的研究是有必要的. 最近, Liang 等 (2007) 讨论了部分线性变量含误差模型在因变量缺失时的估计问题. 然而, 该文章的估计方法只是基于完整观测 (complete-case) 数据, 即将存在缺失因变量的观测数据弃之不用, 没有讨论插补技术的使用. 为了更进一步研究因变量缺失下的变量含误差模型, 本书将重点考虑模型 (6.1) 在因变量 Y 存在缺失和自变量 $\boldsymbol{Z}$ 测量含误差下的估计问题. 为了描述因变量的缺失, 引入新变量 δ, $\delta=1$ 表示 Y 值可以被观测到, 而 $\delta=0$ 表示 Y 存在缺失. 对于自变量 $\boldsymbol{Z}$, 得不到其真实值, 测量得到的是 $\boldsymbol{V}$, 二者有如下关系

$$\boldsymbol{V}=\boldsymbol{Z}+\boldsymbol{\xi}. \tag{6.2}$$

测量误差 $\boldsymbol{\xi}$ 独立于 $(\boldsymbol{X}^{\mathrm{T}},\boldsymbol{Z}^{\mathrm{T}},U,\delta)^{\mathrm{T}}$, 其均值为 $\mathbf{0}$, 协方差阵为 $\boldsymbol{\Sigma}_{\boldsymbol{\xi}}$. 为了模型的可识别性, 本章假设 $\boldsymbol{\Sigma}_{\boldsymbol{\xi}}$ 已知. 如果该协方差阵未知, 可以利用 $\boldsymbol{V}$ 的重复观测数据得到其估计值. 本章假定如下的缺失机制

$$\mathrm{pr}(\delta=1|Y,\boldsymbol{X},\boldsymbol{Z},U)=\mathrm{pr}(\delta=1|\boldsymbol{X},\boldsymbol{Z},U)=\pi(\boldsymbol{X},\boldsymbol{Z},U).$$

值得注意的是, 如果 $\boldsymbol{Z}$ 可以被精确观测, 那么因变量 Y 是随机缺失 (missing at random). 但是, 本书中 $\boldsymbol{Z}$ 的精确值不能得到, 又没有对缺失概率做进一步假定, 因此, Y 不是随机缺失.

就作者所知, 目前研究变量含误差模型在因变量缺失下估计的结果还较少. 本章的主要内容是构造参数分量 $\boldsymbol{\beta}$ 以及非参数分量 $\boldsymbol{\alpha}(\cdot)$ 和因变量均值 $E(Y)$ 的估计. 将分别利用完整数据以及插补技术和替代技术提出参数分量和非参数分量的多种估计量, 并研究 $E(Y)$ 的估计问题.

6.2 参数分量和非参数分量的几类估计

6.2.1 完整观测数据估计方法

对于回归模型因变量存在缺失的情况, 一个简单直接的处理方法就是完整数据

(complete-case data) 方法, 即只将因变量不存在缺失的那些完整观测值用于统计推断, 也就是只利用对应于 $\delta_i = 1$ 的观测数据. 下面将基于校正的 profile 最小二乘方法构造参数分量和非参数分量的估计. 为了方便起见, 首先假定自变量 $\boldsymbol{Z}$ 能精确观测. 假设 $\{Y_i, \delta_i, \boldsymbol{X}_i, \boldsymbol{Z}_i, U_i\}_{i=1}^n$ 为来自模型 (6.1) 的观测数据, 则有

$$\delta_i Y_i = \delta_i \boldsymbol{Z}_i^{\mathrm{T}} \boldsymbol{\beta} + \delta_i \boldsymbol{X}_i^{\mathrm{T}} \boldsymbol{\alpha}(U_i) + \delta_i \varepsilon_i, \quad i = 1, 2, \cdots, n. \tag{6.3}$$

假定 $\boldsymbol{\beta}$ 已知, 则模型 (6.3) 转化为如下形式的变系数模型

$$\delta_i (Y_i - \boldsymbol{Z}_i^{\mathrm{T}} \boldsymbol{\beta}) = \delta_i \boldsymbol{\alpha}_1(U_i) X_{i1} + \cdots + \delta_i \boldsymbol{\alpha}_p(U_i) X_{ip} + \delta_i \varepsilon_i. \tag{6.4}$$

利用基于局部线性光滑的局部加权最小二乘法来估计未知系数函数. 给定 u_0 邻域内的一点 u, 对 $\alpha_j(u)$ 利用 Taylor 展开式有

$$\alpha_j(u) \approx \alpha_j(u_0) + \alpha_j'(u_0)(u - u_0) \equiv a_j + b_j(u - u_0), \quad j = 1, 2, \cdots, p.$$

从而可通过针对 $a_j(u_0), a_j'(u_0)$ 极小化

$$\sum_{i=1}^n \left[\left(Y_i - \boldsymbol{Z}_i^{\mathrm{T}} \boldsymbol{\beta} \right) - \sum_{j=1}^p \{a_j + b_j(U_i - u_0)\} X_{ij} \right]^2 K_{h_1}(U_i - u_0) \delta_i \tag{6.5}$$

来得到它们的估计 $(\hat{\alpha}_j(u_0), \hat{\alpha}_j'(u_0), j = 1, 2, \cdots, p)$, 其中 $K_{h_1}(\cdot) = K(\cdot/h_1)/h_1$, K 是核函数, h_1 是窗宽.

基于问题 (6.5), 由广义最小二乘法可得

$$\overline{\boldsymbol{\alpha}}_c(u) = (\boldsymbol{I}_p \ \ \boldsymbol{0}_p) \{\boldsymbol{D}_{u_0}^{\mathrm{T}} \boldsymbol{W}_{u_0}^{\delta} \boldsymbol{D}_{u_0}\}^{-1} \boldsymbol{D}_{u_0}^{\mathrm{T}} \boldsymbol{W}_{u_0}^{\delta} (\boldsymbol{Y} - \boldsymbol{Z}\boldsymbol{\beta}), \tag{6.6}$$

其中

$$\boldsymbol{Z} = \begin{pmatrix} \boldsymbol{Z}_1^{\mathrm{T}} \\ \boldsymbol{Z}_2^{\mathrm{T}} \\ \vdots \\ \boldsymbol{Z}_n^{\mathrm{T}} \end{pmatrix} = \begin{pmatrix} Z_{11} & \cdots & Z_{1q} \\ Z_{21} & \cdots & Z_{2q} \\ \vdots & & \vdots \\ Z_{n1} & \cdots & Z_{nq} \end{pmatrix},$$

$$\boldsymbol{D}_{u_0} = \begin{pmatrix} \boldsymbol{X}_1^{\mathrm{T}} & \dfrac{U_1 - u_0}{h_1} \boldsymbol{X}_1^{\mathrm{T}} \\ \boldsymbol{X}_2^{\mathrm{T}} & \dfrac{U_2 - u_0}{h_1} \boldsymbol{X}_2^{\mathrm{T}} \\ \vdots & \vdots \\ \boldsymbol{X}_n^{\mathrm{T}} & \dfrac{U_n - u_0}{h_1} \boldsymbol{X}_n^{\mathrm{T}} \end{pmatrix}, \quad \boldsymbol{Y} = \begin{pmatrix} Y_1 \\ Y_2 \\ \vdots \\ Y_n \end{pmatrix},$$

和

$$\boldsymbol{W}_{u_0}^{\delta} = \mathrm{diag}(K_{h_1}(U_1 - u_0)\delta_1, K_{h_1}(U_2 - u_0)\delta_2, \cdots, K_{h_1}(U_n - u_0)\delta_n).$$

用 $\overline{\boldsymbol{\alpha}}_c(U_i)$ 代替模型 (6.3) 中的 $\boldsymbol{\alpha}(U_i)$, 经过简单的计算有

$$\delta_i(Y_i - \hat{Y}_i) = \delta_i(\boldsymbol{Z}_i - \hat{\boldsymbol{Z}}_i)^{\mathrm{T}}\boldsymbol{\beta} + \delta_i\varepsilon_i, \tag{6.7}$$

其中 $\hat{\boldsymbol{Y}} = (\hat{Y}_1, \cdots, \hat{Y}_n)^{\mathrm{T}} = \boldsymbol{S}_c\boldsymbol{Y}$ 和 $\hat{\boldsymbol{Z}} = (\hat{\boldsymbol{Z}}_1, \cdots, \hat{\boldsymbol{Z}}_n)^{\mathrm{T}} = \boldsymbol{S}_c\boldsymbol{Z}$,

$$\boldsymbol{S}_c = \begin{pmatrix} (\boldsymbol{X}_1^{\mathrm{T}}\ \boldsymbol{0})\{\boldsymbol{D}_{u_1}^{\mathrm{T}}\boldsymbol{W}_{u_1}^{\delta}\boldsymbol{D}_{u_1}\}^{-1}\boldsymbol{D}_{u_1}^{\mathrm{T}}\boldsymbol{W}_{u_1}^{\delta} \\ (\boldsymbol{X}_2^{\mathrm{T}}\ \boldsymbol{0})\{\boldsymbol{D}_{u_2}^{\mathrm{T}}\boldsymbol{W}_{u_2}^{\delta}\boldsymbol{D}_{u_2}\}^{-1}\boldsymbol{D}_{u_2}^{\mathrm{T}}\boldsymbol{W}_{u_2}^{\delta} \\ \vdots \\ (\boldsymbol{X}_n^{\mathrm{T}}\ \boldsymbol{0})\{\boldsymbol{D}_{u_n}^{\mathrm{T}}\boldsymbol{W}_{u_n}^{\delta}\boldsymbol{D}_{u_n}\}^{-1}\boldsymbol{D}_{u_n}^{\mathrm{T}}\boldsymbol{W}_{u_n}^{\delta} \end{pmatrix}. \tag{6.8}$$

利用最小二乘法估计模型 (6.7), 可得 $\boldsymbol{\beta}$ 基于完整观测数据的 profile 最小二乘估计

$$\overline{\boldsymbol{\beta}} = \left\{\sum_{i=1}^{n}\delta_i(\boldsymbol{Z}_i - \hat{\boldsymbol{Z}}_i)(\boldsymbol{Z}_i - \hat{\boldsymbol{Z}}_i)^{\mathrm{T}}\right\}^{-1}\sum_{i=1}^{n}\delta_i(\boldsymbol{Z}_i - \hat{\boldsymbol{Z}}_i)(Y_i - \hat{Y}_i).$$

然而, 上面的估计假定了数据 $\boldsymbol{Z}_i$ 可以得到. 如果简单的用 $\boldsymbol{V}_i$ 代替 $\boldsymbol{Z}_i$ 而不考虑测量误差的问题, 很容易可以证明这样得到的估计是不相合的. 为了解决这一问题, 构造 $\boldsymbol{\beta}$ 的校正估计如下

$$\hat{\boldsymbol{\beta}}_c = \left\{\sum_{i=1}^{n}\delta_i\left[(\boldsymbol{V}_i - \hat{\boldsymbol{V}}_i)(\boldsymbol{V}_i - \hat{\boldsymbol{V}}_i)^{\mathrm{T}} - \boldsymbol{\Sigma}_{\boldsymbol{\xi}}\right]\right\}^{-1}\sum_{i=1}^{n}\delta_i(\boldsymbol{V}_i - \hat{\boldsymbol{V}}_i)(Y_i - \hat{Y}_i), \tag{6.9}$$

其中 $(\hat{\boldsymbol{V}}_1, \cdots, \hat{\boldsymbol{V}}_n)^{\mathrm{T}} = \hat{\boldsymbol{V}} = \boldsymbol{S}_c\boldsymbol{V}$, $\boldsymbol{V} = (\boldsymbol{V}_1, \cdots, \boldsymbol{V}_n)^{\mathrm{T}}$. 接下来, 定义 $\boldsymbol{\alpha}(u)$ 的估计为

$$\hat{\boldsymbol{\alpha}}_c(u) = (\boldsymbol{I}_p\ \ \boldsymbol{0}_p)\{\boldsymbol{D}_u^{\mathrm{T}}\boldsymbol{W}_u^{\delta}\boldsymbol{D}_u\}^{-1}\boldsymbol{D}_u^{\mathrm{T}}\boldsymbol{W}_u^{\delta}(\boldsymbol{Y} - \boldsymbol{V}\hat{\boldsymbol{\beta}}). \tag{6.10}$$

下面给出 $\hat{\boldsymbol{\beta}}_c$ 和 $\hat{\boldsymbol{\alpha}}_c(u)$ 的渐近性质.

定理 6.1 如果 6.4 节的条件成立, $\boldsymbol{\beta}$ 基于完整观测数据的校正 profile 最小二乘估计是渐近正态的,

$$\sqrt{n}(\hat{\boldsymbol{\beta}}_c - \boldsymbol{\beta}) \xrightarrow{d} N(\boldsymbol{0}, \Sigma_1^{-1}\boldsymbol{\Omega}_1\boldsymbol{\Sigma}_1^{-1}),$$

其中

$$\begin{aligned}\boldsymbol{\Sigma}_1 =& E\left\{\delta[\boldsymbol{Z} - \boldsymbol{\Phi}_c(U)\boldsymbol{\Gamma}_c^{-1}(U)\boldsymbol{X}]^{\otimes 2}\right\},\\ \boldsymbol{\Omega}_1 =& E\left\{\delta[(\varepsilon - \boldsymbol{\xi}^{\mathrm{T}}\boldsymbol{\beta})(\boldsymbol{Z} - \boldsymbol{\Phi}_c(U)\boldsymbol{\Gamma}_c^{-1}(U)\boldsymbol{X})]^{\otimes 2}\right\} + E(\delta\varepsilon^2\xi\xi^{\mathrm{T}})\\ &+E\left\{\delta[(\boldsymbol{\xi}\boldsymbol{\xi}^{\mathrm{T}} - \boldsymbol{\Sigma}_{\boldsymbol{\xi}})\boldsymbol{\beta}]^{\otimes 2}\right\},\end{aligned}$$

$\boldsymbol{\Gamma}_c(U)=E(\delta\boldsymbol{X}\boldsymbol{X}^{\mathrm{T}}|U)$ 和 $\boldsymbol{\Phi}_c(U)=E(\delta\boldsymbol{Z}\boldsymbol{X}^{\mathrm{T}}|U)$, $\boldsymbol{A}^{\otimes 2}$ 表示 $\boldsymbol{A}\boldsymbol{A}^{\mathrm{T}}$.

注　如果因变量不存在缺失 (即 $\pi(\boldsymbol{Z},\boldsymbol{X},U)=1$), ε_i 具有相同的方差 σ^2, 定理 2.1 即为 You 和 Chen(2006) 中的定理 3.1.

为了利用上面的结论进行推断, 有时候需要估计渐近协方差阵 $\boldsymbol{\Sigma}_1^{-1}\boldsymbol{\Omega}_1\boldsymbol{\Sigma}_1^{-1}$. 令

$$\hat{\boldsymbol{\Sigma}}_1=n^{-1}\sum_{i=1}^{n}\delta_i\left\{(\boldsymbol{V}_i-\hat{\boldsymbol{V}}_i)(\boldsymbol{V}_i-\hat{\boldsymbol{V}}_i)^{\mathrm{T}}-\boldsymbol{\Sigma}_\xi\right\},$$

$$\hat{\boldsymbol{\Omega}}_1=n^{-1}\sum_{i=1}^{n}\delta_i\left\{(\boldsymbol{V}_i-\hat{\boldsymbol{V}}_i)[\boldsymbol{Y}_i-\hat{\boldsymbol{Y}}_i-(\boldsymbol{V}_i-\hat{\boldsymbol{V}}_i)^{\mathrm{T}}\hat{\boldsymbol{\beta}}_c]+\boldsymbol{\Sigma}_{\boldsymbol{\xi}}\hat{\boldsymbol{\beta}}_c\right\}^{\otimes 2},$$

从定理 2.1 的证明过程很容易得出 $\hat{\boldsymbol{\Sigma}}_1^{-1}\hat{\boldsymbol{\Omega}}_1\hat{\boldsymbol{\Sigma}}_1^{-1}$ 即为 $\boldsymbol{\Sigma}_1^{-1}\boldsymbol{\Omega}_1\boldsymbol{\Sigma}_1^{-1}$ 的相合估计.

定理 6.2　如果 6.4 节的条件成立, $h_1=cn^{-1/5}$, 其中 c 为一常数, 有

$$\max_{1\leqslant j\leqslant p}\sup_{U\in\Omega}|\hat{\alpha}_{cj}(u)-\alpha_j(u)|=O\{n^{-2/5}(\log n)^{1/2}\},\quad \text{a.s.}$$

前面基于完整观测数据给出了部分线性变系数变量含误差模型在因变量缺失下的估计, Wei 和 Mei (2012) 针对此种情况进一步利用经验似然方法构造了参数分量的区间估计.

6.2.2　插补估计

为了充分利用观测数据的信息, 可以利用 Chu 和 Cheng (1995) 中的插补方法. 该方法的思想是利用单点插补技术补全缺失的因变量, 从而所有的数据都参与后面的推断过程. 具体来讲, 首先用前面介绍的完整观测数据方法分别得到参数分量和非参数分量的估计 $\hat{\boldsymbol{\beta}}_c$ 和 $\hat{\boldsymbol{\alpha}}_c(U_i)$. 如果 $\boldsymbol{Z}$ 可以被精确观测, 有观测数据 $(Y_i^0,\boldsymbol{X}_i,\boldsymbol{Z}_i,U_i)_{i=1}^n$, 其中

$$Y_i^0=\delta_iY_i+(1-\delta_i)\left(\boldsymbol{X}_i^{\mathrm{T}}\hat{\boldsymbol{\alpha}}_c(U_i)+\boldsymbol{Z}_i^{\mathrm{T}}\hat{\boldsymbol{\beta}}_c\right).$$

然而, 由于 $\boldsymbol{Z}_i$ 精确值的不存在, 得不到 Y_i^0 的精确值, 得到的是 $Y_i^*=\delta_iY_i+(1-\delta_i)\left(\boldsymbol{X}_i^{\mathrm{T}}\hat{\boldsymbol{\alpha}}_c(U_i)+\boldsymbol{V}_i^{\mathrm{T}}\hat{\boldsymbol{\beta}}_c\right)$. 基于构造数据 $(Y_i^*,\boldsymbol{X}_i,\boldsymbol{V}_i,U_i)_{i=1}^n$, 有如下因变量和自变量都存在测量误差的部分线性变系数线性模型

$$\begin{cases}Y_i^0=\boldsymbol{X}_i^{\mathrm{T}}\boldsymbol{\alpha}(U_i)+\boldsymbol{Z}_i^{\mathrm{T}}\boldsymbol{\beta}+e_i,\quad i=1,2,\cdots,n,\\ \boldsymbol{V}_i=\boldsymbol{Z}_i+\boldsymbol{\xi}_i,\\ Y_i^*=Y_i^0+(1-\delta_i)\boldsymbol{\xi}_i^{\mathrm{T}}\hat{\boldsymbol{\beta}}_c,\end{cases}\tag{6.11}$$

其中 $e_i=Y_i^0-Y_i+\varepsilon_i$ 为模型误差. 接下来, 基于校正技术, 定义 $\boldsymbol{\beta}$ 的插补估计为

$$\hat{\boldsymbol{\beta}}_I=\left\{\sum_{i=1}^{n}\left[(\boldsymbol{V}_i-\widetilde{\boldsymbol{V}}_i)(\boldsymbol{V}_i-\widetilde{\boldsymbol{V}}_i)^{\mathrm{T}}-\boldsymbol{\Sigma}_{\boldsymbol{\xi}}\right]\right\}^{-1}\left\{\sum_{i=1}^{n}(\boldsymbol{V}_i-\widetilde{\boldsymbol{V}}_i)(Y_i^*-\widetilde{Y}_i^*)\right.$$

$$-\sum_{i=1}^{n}(1-\delta_i)\Sigma_{\boldsymbol{\xi}}\hat{\boldsymbol{\beta}}_c\Bigg\},$$

其中 $\boldsymbol{Y}^*=(Y_1^*,\cdots,Y_n^*)^{\mathrm{T}}$, $\widetilde{\boldsymbol{Y}}^*=(\widetilde{Y}_1^*,\cdots,\widetilde{Y}_n^*)^{\mathrm{T}}=\boldsymbol{S}_I\boldsymbol{Y}^*$, $\widetilde{\boldsymbol{V}}=(\widetilde{\boldsymbol{V}}_1,\cdots,\widetilde{\boldsymbol{V}}_n)^{\mathrm{T}}=\boldsymbol{S}_I\boldsymbol{V}$. $\boldsymbol{S}_I$ 与 $\boldsymbol{S}_c$ 具有相同的形式, 除了将 h_1 变为 h_2 以及 $\boldsymbol{W}_{u_i}^{\delta}$ 变为

$$\boldsymbol{W}_{u_i}=\mathrm{diag}(K_{h_2}(U_1-u_0),K_{h_2}(U_2-u_0),\cdots,K_{h_2}(U_n-u_0)).$$

定义 $\boldsymbol{\alpha}(u)$ 的插补估计为

$$\hat{\boldsymbol{\alpha}}_I(u)=[\hat{\alpha}_{I1}(u),\cdots,\hat{\alpha}_{Ip}(u)]^{\mathrm{T}}=(\boldsymbol{I}_p\ \ \boldsymbol{0}_p)\{\boldsymbol{D}_u^{\mathrm{T}}\boldsymbol{W}_u\boldsymbol{D}_u\}^{-1}\boldsymbol{D}_u^{\mathrm{T}}\boldsymbol{W}_u(\boldsymbol{Y}^*-\boldsymbol{Z}\hat{\boldsymbol{\beta}}_I).$$

下面的定理将给出参数分量和非参数分量插补估计的渐近性质.

定理 6.3 如果 6.4 节的条件成立, $\boldsymbol{\beta}$ 的插补估计是渐近正态的,

$$\sqrt{n}(\hat{\boldsymbol{\beta}}_I-\boldsymbol{\beta})\xrightarrow{d}N(\boldsymbol{0},\boldsymbol{\Sigma}^{-1}\boldsymbol{\Omega}_2\boldsymbol{\Sigma}^{-1}),$$

其中

$$\boldsymbol{\Gamma}(U)=E(\boldsymbol{X}\boldsymbol{X}^{\mathrm{T}}|U),\ \boldsymbol{\Phi}(U)=E(\boldsymbol{Z}\boldsymbol{X}^{\mathrm{T}}|U),$$

$$\boldsymbol{\Sigma}=E\left\{[\boldsymbol{Z}-\boldsymbol{\Phi}(U)\boldsymbol{\Gamma}^{-1}(U)\boldsymbol{X}][\boldsymbol{Z}-\boldsymbol{\Phi}(U)\boldsymbol{\Gamma}^{-1}(U)\boldsymbol{X}]^{\mathrm{T}}\right\},$$

$$\boldsymbol{\Omega}_2=(\boldsymbol{\Sigma}_2+\boldsymbol{\Sigma}_1)\boldsymbol{\Sigma}_1^{-1}\boldsymbol{\Omega}_1\boldsymbol{\Sigma}_1^{-1}(\boldsymbol{\Sigma}_2+\boldsymbol{\Sigma}_1),$$

$$\boldsymbol{\Sigma}_2=E\left\{(1-\delta)[\boldsymbol{Z}-\boldsymbol{\Phi}(U)\boldsymbol{\Gamma}^{-1}(U)\boldsymbol{X}][\boldsymbol{Z}-\boldsymbol{\Phi}_c(U)\boldsymbol{\Gamma}_c^{-1}(U)\boldsymbol{X}]^{\mathrm{T}}\right\}.$$

定理 6.4 如果 6.4 节的条件成立, $h_1=c_1n^{-1/5},h_2=c_2n^{-1/5}$, 其中 c_1,c_2 为常数, 有

$$\max_{1\leqslant j\leqslant p}\sup_{U\in\Omega}|\hat{\alpha}_{Ij}(u)-\alpha_j(u)|=O\{n^{-2/5}(\log n)^{1/2}\},\quad \text{a.s.}$$

6.2.3 Surrogate 估计

Wang 和 Sun (2007) 针对因变量缺失下的部分线性模型提出了一类 Surrogate 估计. 该方法主要是利用因变量的估计值代替其实际观测值. 将该方法推广到变量含误差模型领域里. 基于该方法, 有如下的部分线性变系数变量含误差模型

$$\begin{cases} Y_i^{**}=\boldsymbol{X}_i^{\mathrm{T}}\boldsymbol{\alpha}(U_i)+\boldsymbol{Z}_i^{\mathrm{T}}\boldsymbol{\beta}+e_i,\quad i=1,2,\cdots,n,\\ \boldsymbol{V}_i=\boldsymbol{Z}_i+\boldsymbol{\xi}_i,\end{cases}\tag{6.12}$$

其中 $Y_i^{**}=\boldsymbol{X}_i^{\mathrm{T}}\hat{\boldsymbol{\alpha}}_c(U_i)+\boldsymbol{V}_i^{\mathrm{T}}\hat{\boldsymbol{\beta}}_c$. 定义 $\boldsymbol{\beta}$ 的 surrogate 估计为

$$\hat{\boldsymbol{\beta}}_s=\left\{\sum_{i=1}^{n}(\boldsymbol{V}_i-\overline{\boldsymbol{V}}_i)(\boldsymbol{V}_i-\overline{\boldsymbol{V}}_i)^{\mathrm{T}}-n\boldsymbol{\Sigma}_{\boldsymbol{\xi}}\right\}^{-1}\left\{\sum_{i=1}^{n}(\boldsymbol{V}_i-\overline{\boldsymbol{V}}_i)(Y_i^{**}-\widetilde{Y}_i^{**})-n\boldsymbol{\Sigma}_{\boldsymbol{\xi}}\hat{\boldsymbol{\beta}}_c\right\},$$

其中 $\boldsymbol{Y}^{**}=(Y_1^{**},\cdots,Y_n^{**})^{\mathrm{T}}, \widetilde{\boldsymbol{Y}}^{**}=(\overline{Y}_1^{**},\cdots,\overline{Y}_n^{**})^{\mathrm{T}}=\boldsymbol{S}_s\boldsymbol{Y}^{**}, \overline{\boldsymbol{V}}_i=(\boldsymbol{S}_{si}^{\mathrm{T}}\boldsymbol{V})^{\mathrm{T}}$, $\boldsymbol{S}_s$ 为将 $\boldsymbol{S}_I$ 中的 h_2 替换为 h_3. 相应地, $\boldsymbol{\alpha}(\cdot)$ 的估计定义为

$$\hat{\boldsymbol{\alpha}}_s(u)=[\hat{\alpha}_{s1}(u),\cdots,\hat{\alpha}_{sp}(u)]^{\mathrm{T}}=(\boldsymbol{I}_p\ \ \mathbf{0}_p)\{\boldsymbol{D}_u^{\mathrm{T}}\boldsymbol{W}_u\boldsymbol{D}_u\}^{-1}\boldsymbol{D}_u^{\mathrm{T}}\boldsymbol{W}_u(\boldsymbol{Y}^{**}-\boldsymbol{V}\hat{\boldsymbol{\beta}}_s).$$

定理 6.5 如果 6.4 节的条件成立, $\boldsymbol{\beta}$ 的 surrogate 估计是渐近正态的,

$$\sqrt{n}(\hat{\boldsymbol{\beta}}_s-\boldsymbol{\beta})\xrightarrow{d}N(\mathbf{0},\boldsymbol{\Sigma}^{-1}\boldsymbol{\Omega}_3\Sigma^{-1}),$$

其中 $\boldsymbol{\Omega}_3=\boldsymbol{\Sigma}\boldsymbol{\Sigma}_1^{-1}\boldsymbol{\Omega}_1\boldsymbol{\Sigma}_1^{-1}\boldsymbol{\Sigma}$.

定理 6.6 如果 6.4 节的条件成立, $h_1=b_1n^{-1/5}, h_3=b_2n^{-1/5}$, 其中 b_1,b_2 为常数, 有

$$\max_{1\leqslant j\leqslant p}\sup_{U\in\Omega}|\hat{\alpha}_{sj}(u)-\alpha_j(u)|=O\{n^{-2/5}(\log n)^{1/2}\},\quad \text{a.s.}$$

6.3 因变量均值的估计

对于因变量缺失下的回归模型, 另外一个重要问题是 $E(Y)=\theta$ 的估计. 对于这类问题, Cheng (1990), Wang 等 (2004) 和 Liang 等 (2007) 分别基于非参数回归模型和部分线性模型进行过研究. 类似于 Liang 等 (2007) 的方法, 定义 θ 的两类估计

$$\hat{\theta}_1=n^{-1}\sum_{i=1}^n\delta_iY_i+n^{-1}\sum_{i=1}^n(1-\delta_i)\left\{\boldsymbol{X}_i^{\mathrm{T}}\hat{\boldsymbol{\alpha}}_c(U_i)+\boldsymbol{V}_i^{\mathrm{T}}\hat{\boldsymbol{\beta}}_c\right\},\tag{6.13}$$

和

$$\hat{\theta}_2=n^{-1}\sum_{i=1}^n\left\{\boldsymbol{X}_i^{\mathrm{T}}\hat{\boldsymbol{\alpha}}_c(U_i)+\boldsymbol{V}_i^{\mathrm{T}}\hat{\boldsymbol{\beta}}_c\right\}.\tag{6.14}$$

下面的定理给出 $\hat{\theta}_1$ 和 $\hat{\theta}_2$ 的渐近性质.

定理 6.7 如果 6.4 节的条件成立, $nh_1^4\to 0$, $\hat{\theta}_1$ 是渐近正态的,

$$\sqrt{n}(\hat{\theta}_1-\theta)\xrightarrow{d}N(\mathbf{0},\boldsymbol{\Theta}_1),$$

其中 $\boldsymbol{\Theta}_1=\boldsymbol{\Theta}_{11}+\boldsymbol{\Theta}_{12}$,

$$\check{\boldsymbol{X}}_i^{\delta}=E[(1-\delta_i)\boldsymbol{X}_i^{\mathrm{T}}|U_i]\boldsymbol{\Gamma}_c^{-1}(U_i)\boldsymbol{X}_i, \check{\boldsymbol{V}}_\delta=E(1-\delta_i)(\boldsymbol{V}_i-\boldsymbol{\Phi}_c(U_i)\boldsymbol{\Gamma}_c^{-1}(U_i)\boldsymbol{X}_i),$$

$$\begin{aligned}\boldsymbol{\Theta}_{11}=E\Bigg\{&(1+\check{\boldsymbol{X}}_i^{\delta})\delta_i\varepsilon_i+(1-\delta_i)\boldsymbol{\xi}_i^{\mathrm{T}}\boldsymbol{\beta}\\&+\check{\boldsymbol{V}}_\delta^{\mathrm{T}}\boldsymbol{\Sigma}_1^{-1}\delta_i\left[(\boldsymbol{V}_i-\boldsymbol{\Phi}_c(U_i)\boldsymbol{\Gamma}_c^{-1}(U_i)\boldsymbol{X}_i)(\varepsilon_i-\boldsymbol{\xi}_i^{\mathrm{T}}\boldsymbol{\beta})+\boldsymbol{\Sigma}_{\boldsymbol{\xi}}\boldsymbol{\beta}\right]\Bigg\}^{\otimes 2},\\\boldsymbol{\Theta}_{12}=E(&\boldsymbol{X}^{\mathrm{T}}\boldsymbol{\alpha}(U)+\boldsymbol{Z}^{\mathrm{T}}\boldsymbol{\beta}-\theta)^2.\end{aligned}$$

定理 6.8 如果 6.4 节的条件成立, $nh_1^4 \to 0$, $\hat{\theta}_2$ 是渐近正态的,

$$\sqrt{n}(\hat{\theta}_2 - \theta) \xrightarrow{d} N(\mathbf{0}, \boldsymbol{\Theta}_2),$$

其中 $\boldsymbol{\Theta}_2 = \boldsymbol{\Theta}_{21} + \boldsymbol{\Theta}_{22}$, $\check{\boldsymbol{X}}_i = E(\boldsymbol{X}_i^{\mathrm{T}}|U_i)\boldsymbol{\Gamma}_c^{-1}(U_i)\boldsymbol{X}_i$, $\check{\boldsymbol{V}} = E[\boldsymbol{V}_i - \boldsymbol{\Phi}_c(U_i)\boldsymbol{\Gamma}_c^{-1}(U_i)\boldsymbol{X}_i]$,

$$\begin{aligned}\boldsymbol{\Theta}_{21} = E\Big\{&\check{\boldsymbol{X}}_i\delta_i\varepsilon_i + \boldsymbol{\xi}_i^{\mathrm{T}}\boldsymbol{\beta} + \check{\boldsymbol{V}}^{\mathrm{T}}\boldsymbol{\Sigma}_1^{-1}\delta_i[(\boldsymbol{V}_i - \boldsymbol{\Phi}_c(U_i)\boldsymbol{\Gamma}_c^{-1}(U_i)\boldsymbol{X}_i)(\varepsilon_i - \boldsymbol{\xi}_i^{\mathrm{T}}\boldsymbol{\beta})\\ &+\boldsymbol{\Sigma}_{\boldsymbol{\xi}}\boldsymbol{\beta}]\Big\}^{\otimes 2},\\ \boldsymbol{\Theta}_{12} = E(&\boldsymbol{X}^{\mathrm{T}}\boldsymbol{\alpha}(U) + \boldsymbol{Z}^{\mathrm{T}}\boldsymbol{\beta} - \theta)^2.\end{aligned}$$

6.4 定理的证明

在给出具体证明之前, 下面给出一些假设条件. 这些假设只是为了证明的方便, 并非最弱条件. Fan 和 Huang (2005) 以及 You 和 Chen (2006) 也使用了这些条件.

(A.1) 随机变量 U 具有有界支撑 Ω, 其密度函数 $f(\cdot)$ 在其支撑上满足 Lipschitz 连续, 且不为 0.

(A.2) 对于任一 $U \in \Omega$, 矩阵 $E(\boldsymbol{X}\boldsymbol{X}^{\mathrm{T}}|U)$ 非奇异, $E(\boldsymbol{X}\boldsymbol{X}^{\mathrm{T}}|U)$, $E(\boldsymbol{X}\boldsymbol{X}^{\mathrm{T}}|U)^{-1}$, 且 $E(\boldsymbol{X}\boldsymbol{Z}^{\mathrm{T}}|U)$ 都是 Lipschitz 连续的.

(A.3) 存在 $s > 2$ 使得 $E\|\boldsymbol{X}\|^{2s} < \infty$ 和 $E\|\boldsymbol{Z}\|^{2s} < \infty$, 对于 $\varepsilon < 2 - s^{-1}$ 使得 $n^{2\varepsilon-1}h \to \infty$.

(A.4) $\boldsymbol{\alpha}(\cdot)$ 二阶连续可导.

(A.5) 函数 $K(\cdot)$ 为对称密度函数, 具有紧支撑.

(A.6) $nh^8 \to 0$ 和 $nh^2/(\log n)^2 \to \infty$.

首先给出几个引理. 令 $\mu_i = \int_0^\infty t^i K(t)\mathrm{d}t, \nu_i = \int_0^\infty t^i K^2(t)\mathrm{d}t$, $c_{n_i} = h_i^2 + \left\{\dfrac{\log(1/h_i)}{nh_i}\right\}^{1/2}, i = 1, 2$, 且 $\Delta = \mathrm{diag}(\delta_1, \cdots, \delta_n)$.

引理 6.1 令 $(\boldsymbol{X}_1, \boldsymbol{Y}_1), \cdots, (\boldsymbol{X}_n, \boldsymbol{Y}_n)$ 为独立同分布 (iid) 的随机序列, 其中 $\boldsymbol{Y}_i, i = 1, 2, \cdots, n$ 为一元随机变量, 进一步假定 $E|y|^s < \infty$ 与 $\sup_x \int |y|^s f(x, y)\mathrm{d}y < \infty$, 其中 f 表示 $(\boldsymbol{X}, \boldsymbol{Y})$ 的联合密度. 令 K 为一有界正函数, 并有有界支撑且满足 Lipschitz 条件, 则有

$$\sup_x\left|\frac{1}{n}\sum_{i=1}^n [K_h(\boldsymbol{X}_i - x)\boldsymbol{Y}_i - E\left(K_h(\boldsymbol{X}_i - x)\boldsymbol{Y}_i\right)]\right| = O_p\left(\left\{\frac{\log(1/h)}{nh}\right\}^{1/2}\right),$$

其中对于 $\varepsilon < 1 - s^{-1}$, 有 $n^{2\varepsilon-1}h \to \infty$. 该引理由 Mack 和 Silverman (1982) 即可证得.

引理 6.2　若条件 (A1)~ 条件 (A.5) 成立, 有

$$n^{-1}\sum_{i=1}^{n}\delta_i(\boldsymbol{Z}_i-\hat{\boldsymbol{Z}}_i)(\boldsymbol{Z}_i-\hat{\boldsymbol{Z}}_i)^{\mathrm{T}}\xrightarrow{P}\boldsymbol{\Sigma}_1,$$
$$n^{-1}\sum_{i=1}^{n}(\boldsymbol{Z}_i-\widetilde{\boldsymbol{Z}}_i)(\boldsymbol{Z}_i-\widetilde{\boldsymbol{Z}}_i)^{\mathrm{T}}\xrightarrow{P}\boldsymbol{\Sigma}.$$

该引理的证明类似于 Fan 和 Huang (2005) 中引理 7.2 的证明, 具体过程在此省略.

定理 6.1 的证明　记 $\boldsymbol{\nabla}_n=n^{-1}\sum_{i=1}^{n}\delta_i\left[(\boldsymbol{V}_i-\hat{\boldsymbol{V}}_i)(\boldsymbol{V}_i-\hat{\boldsymbol{V}}_i)^{\mathrm{T}}-\boldsymbol{\Sigma}_{\boldsymbol{\xi}}\right]$. 根据 $\hat{\boldsymbol{\beta}}_c$ 的定义, 有

$$\begin{aligned}\sqrt{n}(\hat{\boldsymbol{\beta}}_c-\boldsymbol{\beta})&=\boldsymbol{\nabla}_n{}^{-1}\frac{1}{\sqrt{n}}\sum_{i=1}^{n}\delta_i\left\{(\boldsymbol{V}_i-\hat{\boldsymbol{V}}_i)[Y_i-\hat{Y}_i-(\boldsymbol{V}_i-\hat{\boldsymbol{V}}_i)^{\mathrm{T}}\boldsymbol{\beta}]+\boldsymbol{\Sigma}_{\boldsymbol{\xi}}\boldsymbol{\beta}\right\}\\&=\boldsymbol{\nabla}_n{}^{-1}\frac{1}{\sqrt{n}}\sum_{i=1}^{n}\delta_i(\boldsymbol{V}_i-\hat{\boldsymbol{V}}_i)(\varepsilon_i-\boldsymbol{\xi}_i^{\mathrm{T}}\boldsymbol{\beta})\\&\quad+\boldsymbol{\nabla}_n{}^{-1}\frac{1}{\sqrt{n}}\sum_{i=1}^{n}\delta_i(\boldsymbol{V}_i-\hat{\boldsymbol{V}}_i)(\boldsymbol{X}_i^{\mathrm{T}}\boldsymbol{\alpha}(U_i)-\boldsymbol{S}_{ci}^{\mathrm{T}}\boldsymbol{M})\\&\quad+\boldsymbol{\nabla}_n{}^{-1}\frac{1}{\sqrt{n}}\sum_{i=1}^{n}\delta_i(\boldsymbol{V}_i-\hat{\boldsymbol{V}}_i)\boldsymbol{S}_{ci}^{\mathrm{T}}(\boldsymbol{\varepsilon}-\boldsymbol{\xi}\boldsymbol{\beta})\\&\quad+\boldsymbol{\nabla}_n{}^{-1}\frac{1}{\sqrt{n}}\sum_{i=1}^{n}\delta_i\boldsymbol{\Sigma}_{\boldsymbol{\xi}}\boldsymbol{\beta}\\&\doteq I_1+I_2+I_3+I_4.\end{aligned}$$

基于引理 6.1, 类似于 Fan 和 Huang (2005) 中定理 4.1 的证明, 很容易得到

$$\begin{aligned}&I_1=\boldsymbol{\nabla}_n{}^{-1}\frac{1}{\sqrt{n}}\sum_{i=1}^{n}\delta_i(\boldsymbol{V}_i-\boldsymbol{\Phi}_c(U_i)\boldsymbol{\Gamma}_c(U_i)\boldsymbol{X}_i)(\varepsilon_i-\boldsymbol{\xi}_i^{\mathrm{T}}\boldsymbol{\beta})+o_p(1),\\&I_2=o_p(1),I_3=o_p(1).\end{aligned}$$

再由引理 6.2, 可得

$$\sqrt{n}(\hat{\boldsymbol{\beta}}_c-\boldsymbol{\beta})=\boldsymbol{\Sigma}_1^{-1}\frac{1}{\sqrt{n}}\sum_{i=1}^{n}\delta_i\left\{(\boldsymbol{V}_i-\boldsymbol{\Phi}_c(U_i)\boldsymbol{\Gamma}_c(U_i)\boldsymbol{X}_i)(\varepsilon_i-\boldsymbol{\xi}_i^{\mathrm{T}}\boldsymbol{\beta})+\boldsymbol{\Sigma}_{\boldsymbol{\xi}}\boldsymbol{\beta}\right\}+o_p(1),$$

则由中心极限定理和 Slutsky 定理, 可完成定理 6.1 的证明.

定理 6.2 的证明　由 $\hat{\boldsymbol{\alpha}}_c(u)$ 的定义, 有

$$\begin{aligned}\hat{\boldsymbol{\alpha}}_c(u)&=(\boldsymbol{I}_p\ \ \boldsymbol{0}_p)\{\boldsymbol{D}_u^{\mathrm{T}}\boldsymbol{W}_u^{\delta}\boldsymbol{D}_u\}^{-1}\boldsymbol{D}_u^{\mathrm{T}}\boldsymbol{W}_u^{\delta}(\boldsymbol{Y}-\boldsymbol{V}\hat{\boldsymbol{\beta}}_c)\\&=(\boldsymbol{I}_p\ \ \boldsymbol{0}_p)\{\boldsymbol{D}_u^{\mathrm{T}}\boldsymbol{W}_u^{\delta}\boldsymbol{D}_u\}^{-1}\boldsymbol{D}_u^{\mathrm{T}}\boldsymbol{W}_u^{\delta}\boldsymbol{M}\\&\quad+(\boldsymbol{I}_p\ \ \boldsymbol{0}_p)\{\boldsymbol{D}_u^{\mathrm{T}}\boldsymbol{W}_u^{\delta}\boldsymbol{D}_u\}^{-1}\boldsymbol{D}_u^{\mathrm{T}}\boldsymbol{W}_u^{\delta}(\boldsymbol{\varepsilon}-\boldsymbol{\xi}\boldsymbol{\beta})\\&\quad+(\boldsymbol{I}_p\ \ \boldsymbol{0}_p)\{\boldsymbol{D}_u^{\mathrm{T}}\boldsymbol{W}_u^{\delta}\boldsymbol{D}_u\}^{-1}\boldsymbol{D}_u^{\mathrm{T}}\boldsymbol{W}_u^{\delta}\boldsymbol{V}(\boldsymbol{\beta}-\hat{\boldsymbol{\beta}}_c).\end{aligned}$$

基于定理 6.1 的结论, 类似于 Xia 和 Li (1999) 中定理 3.1 的证明, 能得到

$$\max_{1\leqslant j\leqslant p}\sup_{U\in\Omega}|\hat{\alpha}_{cj}(u)-\alpha_j(u)|=O\{h_1^2+(\log n/nh_1)^{1/2}\},\quad \text{a.s..}$$

如果 h_1 取 $h_1=cn^{-1/5}$, 其中 c 为常数, 则有

$$\max_{1\leqslant j\leqslant p}\sup_{U\in\Omega}|\hat{\alpha}_{cj}(u)-\alpha_j(u)|=O\{n^{-2/5}(\log n)^{1/2}\},\quad \text{a.s..}$$

类似于定理 6.2 的证明, 可以证明定理 6.4 和定理 6.6, 具体细节在此省略.

定理 6.3 的证明 记 $\boldsymbol{\Lambda}_n=n^{-1}\sum\limits_{i=1}^{n}(\boldsymbol{V}_i-\widetilde{\boldsymbol{V}}_i)(\boldsymbol{V}_i-\widetilde{\boldsymbol{V}}_i)^{\mathrm{T}}-\boldsymbol{\Sigma}_{\boldsymbol{\xi}}$. 由 $\hat{\boldsymbol{\beta}}_I$ 的定义, 有

$$\begin{aligned}
\sqrt{n}(\hat{\boldsymbol{\beta}}_I-\boldsymbol{\beta})=&\Lambda_n^{-1}\frac{1}{\sqrt{n}}\sum_{i=1}^{n}(\boldsymbol{V}_i-\widetilde{\boldsymbol{V}}_i)\left\{Y_i^*-\widetilde{Y}_i^*-(\boldsymbol{V}_i-\widetilde{\boldsymbol{V}}_i)^{\mathrm{T}}\boldsymbol{\beta}\right\}\\
&+\boldsymbol{\Lambda}_n^{-1}\left\{\sqrt{n}\boldsymbol{\Sigma}_{\boldsymbol{\xi}}\boldsymbol{\beta}-\frac{1}{\sqrt{n}}\sum_{i=1}^{n}(1-\delta_i)\boldsymbol{\Sigma}_{\boldsymbol{\xi}}\hat{\boldsymbol{\beta}}_c\right\}\\
=&\boldsymbol{\Lambda}_n^{-1}\frac{1}{\sqrt{n}}\sum_{i=1}^{n}(\boldsymbol{V}_i-\widetilde{\boldsymbol{V}}_i)\delta_i(\varepsilon_i-\boldsymbol{\xi}_i^{\mathrm{T}}\boldsymbol{\beta})\\
&+\boldsymbol{\Lambda}_n^{-1}\frac{1}{\sqrt{n}}\sum_{i=1}^{n}(1-\delta_i)(\boldsymbol{V}_i-\widetilde{\boldsymbol{V}}_i)(\boldsymbol{V}_i-\hat{\boldsymbol{V}}_i)^{\mathrm{T}}(\hat{\boldsymbol{\beta}}_c-\boldsymbol{\beta})\\
&+\boldsymbol{\Lambda}_n^{-1}\frac{1}{\sqrt{n}}\sum_{i=1}^{n}(1-\delta_i)(\boldsymbol{Z}_i-\widetilde{\boldsymbol{Z}}_i)\boldsymbol{S}_{ci}^{\mathrm{T}}(\varepsilon-\boldsymbol{\xi}\boldsymbol{\beta})\\
&+\boldsymbol{\Lambda}_n^{-1}\frac{1}{\sqrt{n}}\sum_{i=1}^{n}(\boldsymbol{V}_i-\widetilde{\boldsymbol{V}}_i)(\boldsymbol{X}_i^{\mathrm{T}}\boldsymbol{\alpha}(U_i)+\widetilde{\boldsymbol{V}}_i^{\mathrm{T}}\boldsymbol{\beta}-\boldsymbol{S}_{Ii}^{\mathrm{T}}Y^*)\\
&+\boldsymbol{\Lambda}_n^{-1}\frac{1}{\sqrt{n}}\sum_{i=1}^{n}(1-\delta_i)(\boldsymbol{V}_i-\widetilde{\boldsymbol{V}}_i)(\boldsymbol{S}_{ci}^{\mathrm{T}}\boldsymbol{M}-\boldsymbol{X}_i^{\mathrm{T}}\boldsymbol{\alpha}(U_i))\\
&+\boldsymbol{\Lambda}_n^{-1}\left\{\sqrt{n}\boldsymbol{\Sigma}_{\boldsymbol{\xi}}\boldsymbol{\beta}-\frac{1}{\sqrt{n}}\sum_{i=1}^{n}(1-\delta_i)\boldsymbol{\Sigma}_{\boldsymbol{\xi}}\hat{\boldsymbol{\beta}}_c\right\}\\
\doteq&\Lambda_n^{-1}(I_1+I_2+I_3+I_4+I_5+I_6).
\end{aligned}\tag{6.15}$$

由引理 6.1 以及 $E(\boldsymbol{X}_i\boldsymbol{V}_i^{\mathrm{T}}|U_i)=E(\boldsymbol{X}_i\boldsymbol{Z}_i^{\mathrm{T}}|U_i)$, 易证

$$\widetilde{\boldsymbol{V}}_i=\boldsymbol{\Phi}(U_i)\boldsymbol{\Gamma}(U_i)\boldsymbol{X}_i+O_p(c_{n_2}),\quad \hat{\boldsymbol{V}}_i=\boldsymbol{\Phi}_c(U_i)\boldsymbol{\Gamma}_c(U_i)\boldsymbol{X}_i+O_p(c_{n_1}).\tag{6.16}$$

经过简单的计算, 有

$$I_1=\frac{1}{\sqrt{n}}\sum_{i=1}^{n}(\boldsymbol{V}_i-\boldsymbol{\Phi}(U_i)\boldsymbol{\Gamma}(U_i)\boldsymbol{X}_i)\,\delta_i(\varepsilon_i-\boldsymbol{\xi}_i^{\mathrm{T}}\boldsymbol{\beta})+o_p(1).\tag{6.17}$$

对于 I_2, 由定理 6.1 以及大数定律, 有

$$\begin{aligned}
I_2 =& \left\{ \frac{1}{n}\sum_{i=1}^{n}(1-\delta_i)(\boldsymbol{V}_i - \boldsymbol{\Phi}(U_i)\boldsymbol{\Gamma}(U_i)\boldsymbol{X}_i)(\boldsymbol{V}_i - \boldsymbol{\Phi}_c(U_i)\boldsymbol{\Gamma}_c(U_i)\boldsymbol{X}_i)^{\mathrm{T}} \right\} \sqrt{n}(\hat{\boldsymbol{\beta}}_c - \boldsymbol{\beta}) \\
&+ o_p(1) \\
=& \left\{ \frac{1}{n}\sum_{i=1}^{n}(1-\delta_i)\left[(\boldsymbol{V}_i - \boldsymbol{\Phi}(U_i)\boldsymbol{\Gamma}(U_i)\boldsymbol{X}_i)(\boldsymbol{V}_i - \boldsymbol{\Phi}_c(U_i)\boldsymbol{\Gamma}_c(U_i)\boldsymbol{X}_i)^{\mathrm{T}} - \boldsymbol{\Sigma}_{\boldsymbol{\xi}}\right] \right\} \\
&\sqrt{n}(\hat{\boldsymbol{\beta}}_c - \boldsymbol{\beta}) \\
&+ \frac{1}{\sqrt{n}}\sum_{i=1}^{n}(1-\delta_i)\boldsymbol{\Sigma}_{\boldsymbol{\xi}}(\hat{\boldsymbol{\beta}}_c - \boldsymbol{\beta}) + o_p(1) \\
=& \boldsymbol{\Sigma}_2\boldsymbol{\Sigma}_1^{-1}\frac{1}{\sqrt{n}}\sum_{i=1}^{n}\delta_i\left\{(\boldsymbol{V}_i - \boldsymbol{\Phi}_c(U_i)\boldsymbol{\Gamma}_c(U_i)\boldsymbol{X}_i)(\varepsilon_i - \boldsymbol{\xi}_i^{\mathrm{T}}\boldsymbol{\beta}) + \boldsymbol{\Sigma}_{\boldsymbol{\xi}}\boldsymbol{\beta}\right\} \\
&+ \frac{1}{\sqrt{n}}\sum_{i=1}^{n}(1-\delta_i)\boldsymbol{\Sigma}_{\boldsymbol{\xi}}(\hat{\boldsymbol{\beta}}_c - \boldsymbol{\beta}) + o_p(1).
\end{aligned} \tag{6.18}$$

I_3 可以分解为

$$\begin{aligned}
I_3 =& \frac{1}{\sqrt{n}}\sum_{i=1}^{n}(1-\delta_i)\boldsymbol{V}_i(\boldsymbol{V}_i^{\mathrm{T}}\ \boldsymbol{0})\{\boldsymbol{D}_{u_i}^{\mathrm{T}}\boldsymbol{W}_{u_i}^{\delta}\boldsymbol{D}_{u_i}\}^{-1}\boldsymbol{D}_{u_i}^{\mathrm{T}}\boldsymbol{W}_{u_i}^{\delta}(\varepsilon - \boldsymbol{\xi}\boldsymbol{\beta}) \\
&- \frac{1}{\sqrt{n}}\sum_{i=1}^{n}(1-\delta_i)\widetilde{\boldsymbol{V}}_i(\boldsymbol{X}_i^{\mathrm{T}}\ \boldsymbol{0})\{\boldsymbol{D}_{u_i}^{\mathrm{T}}\boldsymbol{W}_{u_i}^{\delta}\boldsymbol{D}_{u_i}\}^{-1}\boldsymbol{D}_{u_i}^{\mathrm{T}}\boldsymbol{W}_{u_i}^{\delta}(\varepsilon - \boldsymbol{\xi}\boldsymbol{\beta}) \\
\doteq& I_{31} + I_{32}.
\end{aligned} \tag{6.19}$$

由引理 6.1, 有

$$\boldsymbol{D}_{u_i}^{\mathrm{T}}\boldsymbol{W}_{u_i}^{\delta}\boldsymbol{D}_{u_i} = nf(u_i)\boldsymbol{\Gamma}_c(U_i) \otimes \begin{pmatrix} 1 & 0 \\ 0 & \mu_2 \end{pmatrix}\{1 + O_p(c_{n_1})\}. \tag{6.20}$$

基于引理 6.1, 易证

$$\begin{aligned}
I_{31} =& \frac{1}{\sqrt{n}}\sum_{i=1}^{n}(1-\delta_i)\boldsymbol{V}_i\boldsymbol{X}_i^{\mathrm{T}}(nf(u_i)\boldsymbol{\Gamma}_c(U_i))^{-1}\sum_{j=1}^{n}K_{h_1}(U_j - U_i)X_j\delta_j(\varepsilon_j - \boldsymbol{\xi}_j^{\mathrm{T}}\boldsymbol{\beta}) + o_p(1) \\
=& \frac{1}{\sqrt{n}}\sum_{j=1}^{n}\left\{\sum_{i=1}^{n}(1-\delta_i)\boldsymbol{V}_i\boldsymbol{X}_i^{\mathrm{T}}(nf(u_i)\boldsymbol{\Gamma}_c(U_i))^{-1}K_{h_1}(U_j - U_i)\right\} \\
&X_j\delta_j(\varepsilon_j - \boldsymbol{\xi}_j^{\mathrm{T}}\boldsymbol{\beta}) + o_p(1) \\
=& \frac{1}{\sqrt{n}}\sum_{i=1}^{n}\boldsymbol{\Phi}(U_i)\boldsymbol{\Gamma}_c(U_i)\boldsymbol{X}_i\delta_i(\varepsilon_i - \boldsymbol{e}_i^{\mathrm{T}}\boldsymbol{\beta}) - \sum_{i=1}^{n}\boldsymbol{\Phi}_c(U_i)\boldsymbol{\Gamma}_c(U_i)\boldsymbol{X}_i\delta_i(\varepsilon_i - \boldsymbol{\xi}_i^{\mathrm{T}}\boldsymbol{\beta}) + o_p(1).
\end{aligned} \tag{6.21}$$

类似于式 (6.21) 的证明, 有

$$I_{32}=\frac{1}{\sqrt{n}}\sum_{i=1}^{n}\boldsymbol{\Phi}(U_i)\boldsymbol{\Gamma}_c(U_i)\boldsymbol{X}_i\delta_i(\varepsilon_i-\boldsymbol{\xi}_i^{\mathrm{T}}\boldsymbol{\beta})-\sum_{i=1}^{n}\boldsymbol{\Phi}(U_i)\boldsymbol{\Gamma}(U_i)\boldsymbol{X}_i\delta_i(\varepsilon_i-\boldsymbol{\xi}_i^{\mathrm{T}}\boldsymbol{\beta})+o_p(1). \tag{6.22}$$

由式 (6.18)~ 式 (6.22), 可得

$$I_3=\frac{1}{\sqrt{n}}\sum_{i=1}^{n}\{\boldsymbol{\Phi}(U_i)\boldsymbol{\Gamma}(U_i)\boldsymbol{X}_i-\boldsymbol{\Phi}_c(U_i)\boldsymbol{\Gamma}_c(U_i)\boldsymbol{\xi}X_i\}\,\delta_i(\varepsilon_i-\boldsymbol{\xi}_i^{\mathrm{T}}\boldsymbol{\beta})+o_p(1). \tag{6.23}$$

对于 I_4, 有

$$\begin{aligned}I_4=&\frac{1}{\sqrt{n}}\sum_{i=1}^{n}(\boldsymbol{V}_i-\widetilde{\boldsymbol{V}}_i)(\boldsymbol{X}_i^{\mathrm{T}}\boldsymbol{\alpha}(U_i)-\boldsymbol{S}_{Ii}^{\mathrm{T}}\boldsymbol{M})\\&-\frac{1}{\sqrt{n}}\sum_{i=1}^{n}(\boldsymbol{V}_i-\widetilde{\boldsymbol{V}}_i)\boldsymbol{S}_{Ii}^{\mathrm{T}}\Delta(\varepsilon-\boldsymbol{\xi}\boldsymbol{\beta})\\&-\frac{1}{\sqrt{n}}\sum_{i=1}^{n}(\boldsymbol{V}_i-\widetilde{\boldsymbol{V}}_i)\widetilde{\boldsymbol{V}}_i^{\mathrm{T}}\boldsymbol{S}_{Ii}^{\mathrm{T}}(\boldsymbol{I}-\Delta)\boldsymbol{V}(\hat{\boldsymbol{\beta}}_c-\boldsymbol{\beta})\\&-\frac{1}{\sqrt{n}}\sum_{i=1}^{n}(\boldsymbol{V}_i-\widetilde{\boldsymbol{V}}_i)\boldsymbol{S}_{Ii}^{\mathrm{T}}(\boldsymbol{I}-\Delta)(\hat{\boldsymbol{M}}_c-\boldsymbol{M})\\\doteq&\, I_{41}+I_{42}+I_{43}+I_{44}.\end{aligned}$$

其中 $\hat{\boldsymbol{M}}_c=[\boldsymbol{X}_1^{\mathrm{T}}\hat{\boldsymbol{\alpha}}_c(U_1),\cdots,\boldsymbol{X}_n^{\mathrm{T}}\hat{\boldsymbol{\alpha}}_c(U_n)]^{\mathrm{T}}$. 由引理 6.1 可得

$$I_{41}=o_p(1),\quad I_{42}=o_p(1).$$

此外有

$$\frac{1}{n}\sum_{i=1}^{n}(\boldsymbol{V}_i-\widetilde{\boldsymbol{V}}_i)\widetilde{\boldsymbol{V}}_i^{\mathrm{T}}=o_p(1),\quad \boldsymbol{\beta}-\hat{\boldsymbol{\beta}}_c=O_p(n^{-1/2}),$$

基于以上结论, 有 $I_{43}=o_p(1)$.

注意到如下结论

$$\frac{1}{n}\sum_{i=1}^{n}(\boldsymbol{V}_i-\widetilde{\boldsymbol{V}}_i)\boldsymbol{S}_{Ii}^{\mathrm{T}}\boldsymbol{X}=o_p(1),$$

再加上定理 6.2, 易证 $I_{44}=o_p(1)$.

由以上结论, 有

$$I_4=o_p(1). \tag{6.24}$$

由引理 6.1, 可得

$$I_5=O_p(\sqrt{n}c_{n_1}^2). \tag{6.25}$$

由式 (6.17)(6.18)(6.23)(6.24) 和式 (6.25), 有

$$\sqrt{n}(\hat{\boldsymbol{\beta}}_I-\boldsymbol{\beta})=\boldsymbol{\Sigma}^{-1}(\boldsymbol{\Sigma}_1+\boldsymbol{\Sigma}_2)\boldsymbol{\Sigma}_1^{-1}\frac{1}{\sqrt{n}}\sum_{i=1}^{n}\delta_i\Big\{[\boldsymbol{V}_i-\boldsymbol{\Phi}_c(U_i)\boldsymbol{\Gamma}_c(U_i)\boldsymbol{X}_i](\varepsilon_i-\boldsymbol{\xi}_i^{\mathrm{T}}\boldsymbol{\beta})+\boldsymbol{\Sigma}_{\boldsymbol{\xi}}\boldsymbol{\beta}\Big\}+o_p(1),$$

则由中心极限定理和 Slutsky 定理, 可以完成定理 6.3 的证明.

定理 6.5 的证明 记 $\Pi_n=n^{-1}\boldsymbol{V}^{\mathrm{T}}(\boldsymbol{I}-\boldsymbol{S}_s)^{\mathrm{T}}(\boldsymbol{I}-\boldsymbol{S}_s)\boldsymbol{V}-\boldsymbol{\Sigma}_{\boldsymbol{\xi}}$. 由 $\hat{\boldsymbol{\beta}}_s$ 的定义, 有

$$\begin{aligned}\sqrt{n}(\hat{\boldsymbol{\beta}}_s-\boldsymbol{\beta})&=\prod_n^{-1}\frac{1}{\sqrt{n}}\Big\{\boldsymbol{V}^{\mathrm{T}}(\boldsymbol{I}-\boldsymbol{S}_s)^{\mathrm{T}}[\boldsymbol{Y}^{**}-\boldsymbol{S}_s\boldsymbol{Y}^{**}-(\boldsymbol{I}-\boldsymbol{S}_s)\boldsymbol{V}\boldsymbol{\beta}]+n\boldsymbol{\Sigma}_{\boldsymbol{\xi}}(\boldsymbol{\beta}-\hat{\boldsymbol{\beta}}_c)\Big\}\\&=\prod_n^{-1}\frac{1}{\sqrt{n}}\Big\{\boldsymbol{V}^{\mathrm{T}}(\boldsymbol{I}-\boldsymbol{S}_s)^{\mathrm{T}}\boldsymbol{V}-n\boldsymbol{\Sigma}_{\boldsymbol{\xi}}\Big\}(\hat{\boldsymbol{\beta}}_c-\boldsymbol{\beta})\\&\quad+\prod_n^{-1}\frac{1}{\sqrt{n}}\boldsymbol{V}^{\mathrm{T}}(\boldsymbol{I}-\boldsymbol{S}_s)^{\mathrm{T}}(\boldsymbol{I}-\boldsymbol{S}_s)(\hat{\boldsymbol{M}}_c-\boldsymbol{S}_s\boldsymbol{Y}^{**}+\boldsymbol{S}_s\boldsymbol{V}\boldsymbol{\beta})\\&\doteq\prod_n^{-1}(I_1+I_2).\end{aligned}\tag{6.26}$$

由引理 6.1, 有

$$\frac{1}{n}\boldsymbol{V}^{\mathrm{T}}(\boldsymbol{I}-\boldsymbol{S}_s)^{\mathrm{T}}\boldsymbol{V}=\boldsymbol{\Sigma}_{\boldsymbol{\xi}}+\boldsymbol{\Sigma}+o_p(1).$$

从而可以得到

$$I_1=\boldsymbol{\Sigma}\boldsymbol{\Sigma}_1^{-1}\frac{1}{\sqrt{n}}\sum_{i=1}^{n}\delta_i\Big\{[\boldsymbol{V}_i-\boldsymbol{\Phi}_c(U_i)\boldsymbol{\Gamma}_c(U_i)\boldsymbol{X}_i](\varepsilon_i-\boldsymbol{\xi}_i^{\mathrm{T}}\boldsymbol{\beta})+\boldsymbol{\Sigma}_{\boldsymbol{\xi}}\boldsymbol{\beta}\Big\}+o_p(1).$$

类似于定理 6.3 的证明, 易证

$$I_2=o_p(1),$$

则由中心极限定理和 Slutsky 定理, 可以完成定理 6.5 的证明.

定理 6.7 的证明 由 $\hat{\theta}_1$ 的定义, 有

$$\begin{aligned}\hat{\theta}_1&=n^{-1}\sum_{i=1}^{n}\delta_iY_i+n^{-1}\sum_{i=1}^{n}(1-\delta_i)\Big\{\boldsymbol{X}_i^{\mathrm{T}}\hat{\boldsymbol{\alpha}}_c(U_i)+\boldsymbol{V}_i^{\mathrm{T}}\hat{\boldsymbol{\beta}}_c\Big\},\\&=n^{-1}\sum_{i=1}^{n}\delta_i\varepsilon_i+n^{-1}\sum_{i=1}^{n}(1-\delta_i)\boldsymbol{\xi}_i^{\mathrm{T}}\boldsymbol{\beta}+n^{-1}\sum_{i=1}^{n}\Big\{\boldsymbol{X}_i^{\mathrm{T}}\boldsymbol{\alpha}(U_i)+\boldsymbol{Z}_i^{\mathrm{T}}\boldsymbol{\beta}\Big\}\end{aligned}$$

$$
\begin{aligned}
&+n^{-1}\sum_{i=1}^{n}(1-\delta_i)\boldsymbol{V}_i^{\mathrm{T}}(\hat{\boldsymbol{\beta}}_c-\boldsymbol{\beta})\\
&+n^{-1}\sum_{i=1}^{n}(1-\delta_i)(\boldsymbol{X}_i^{\mathrm{T}}\quad \mathbf{0})\{\boldsymbol{D}_{u_i}^{\mathrm{T}}\boldsymbol{W}_{u_i}^{\delta}\boldsymbol{D}_{u_i}\}^{-1}\boldsymbol{D}_{u_i}^{\mathrm{T}}\boldsymbol{W}_{u_i}^{\delta}\boldsymbol{Z}(\boldsymbol{\beta}-\hat{\boldsymbol{\beta}}_c)\\
&+n^{-1}\sum_{i=1}^{n}(1-\delta_i)\left\{(\boldsymbol{X}_i^{\mathrm{T}}\quad \mathbf{0})\{\boldsymbol{D}_{u_i}^{\mathrm{T}}\boldsymbol{W}_{u_i}^{\delta}\boldsymbol{D}_{u_i}\}^{-1}\boldsymbol{D}_{u_i}^{\mathrm{T}}\boldsymbol{W}_{u_i}^{\delta}\boldsymbol{M}-\boldsymbol{X}_i^{\mathrm{T}}\boldsymbol{\alpha}(U_i)\right\}\\
&+n^{-1}\sum_{i=1}^{n}(1-\delta_i)(\boldsymbol{X}_i^{\mathrm{T}}\quad \mathbf{0})\{\boldsymbol{D}_{u_i}^{\mathrm{T}}\boldsymbol{W}_{u_i}^{\delta}\boldsymbol{D}_{u_i}\}^{-1}\boldsymbol{D}_{u_i}^{\mathrm{T}}\boldsymbol{W}_{u_i}^{\delta}\varepsilon\\
\doteq&\sum_{i=1}^{7}I_i.
\end{aligned}
$$

由引理 6.1, 有

$$
I_2+I_3=n^{-1}\sum_{i=1}^{n}(1-\delta_i)[\boldsymbol{V}_i-\boldsymbol{\Phi}_c(U_i)\boldsymbol{\Gamma}_c^{-1}(U_i)\boldsymbol{X}_i)]^{\mathrm{T}}(\hat{\boldsymbol{\beta}}_c-\boldsymbol{\beta})+o_p(n^{-1/2}),I_4=O_p(h_1^2).
$$

对于 I_5, 有

$$
\begin{aligned}
I_5=&n^{-1}\sum_{i=1}^{n}(1-\delta_i)\boldsymbol{X}_i^{\mathrm{T}}(nf(u_i)\boldsymbol{\Gamma}_c(U_i))^{-1}\sum_{j=1}^{n}K_{h_1}(U_j-U_i)X_j\delta_j\varepsilon_j+o_p(n^{-1/2})\\
=&\sum_{j=1}^{n}\left\{\sum_{i=1}^{n}(1-\delta_i)\boldsymbol{X}_i^{\mathrm{T}}(nf(u_i)\boldsymbol{\Gamma}_c(U_i))^{-1}K_{h_1}(U_j-U_i)\right\}X_j\delta_j\varepsilon_j+o_p(n^{-1/2})\\
=&\sum_{j=1}^{n}\check{\boldsymbol{X}}_i^{\delta}\delta_j\varepsilon_j+o_p(n^{-1/2}).
\end{aligned}
$$

如果 $nh_1^4\to 0$, 再由定理 2.1, 有

$$
\begin{aligned}
\sqrt{n}(\hat{\theta}_1-\theta)=&\frac{1}{\sqrt{n}}\sum_{i=1}^{n}\left\{(1+\check{\boldsymbol{X}}_i^{\delta})\delta_i\varepsilon_i+(1-\delta_i)\boldsymbol{\xi}_i^{\mathrm{T}}\boldsymbol{\beta}\right\}\\
&+\frac{1}{\sqrt{n}}\sum_{i=1}^{n}\left\{\check{\boldsymbol{V}}_{\delta}^{\mathrm{T}}\boldsymbol{\Sigma}_1^{-1}\delta_i[(\boldsymbol{V}_i-\boldsymbol{\Phi}_c(U_i)\boldsymbol{\Gamma}_c^{-1}(U_i)\boldsymbol{X}_i)(\varepsilon_i-\boldsymbol{\xi}_i^{\mathrm{T}}\boldsymbol{\beta})+\boldsymbol{\Sigma}_{\boldsymbol{\xi}}\boldsymbol{\beta}]\right\}\\
&+\frac{1}{\sqrt{n}}\sum_{i=1}^{n}\left\{\boldsymbol{X}_i^{\mathrm{T}}\boldsymbol{\alpha}(U_i)+\boldsymbol{Z}_i^{\mathrm{T}}\boldsymbol{\beta}-\theta\right\}+o_p(1),
\end{aligned}
$$

则由中心极限定理, 可完成定理 6.7 的证明.

第 7 章　部分线性可加模型的估计与检验

部分线性可加模型作为部分线性模型和可加模型的推广, 受到了广泛的关注. 然后由于其模型本身的复杂性, 相比于部分线性模型和部分线性变系数模型, 该模型得到的研究还较少. 本章主要考虑这类半参数模型的估计和检验问题, 主要内容来自于文献 Wei 和 Liu (2012).

7.1　引　　言

近年来, 可加模型作为一类重要的半参数模型在统计学和计量经济学领域都受到了广泛的研究和应用. 可加模型最早由 Friedman 和 Stuetzle (1981) 提出, 著作 Hastie 和 Tibshirani (1990) 对其进行了详细的讨论, 该书对可加模型的推广起到了重要的作用. 可加模型的一类重要推广形式是部分线性可加模型, 也被成为半参数可加模型. 部分线性可加模型中一部分自变量是线性形式, 而另一部分是可加结构, 可记为如下形式

$$Y_i = \boldsymbol{X}_i^{\mathrm{T}}\boldsymbol{\beta} + m_1(Z_{1i}) + \cdots + m_q(Z_{qi}) + \varepsilon_i, \quad i = 1, 2, \cdots, n, \tag{7.1}$$

其中 Y_i 是因变量观测值, $\boldsymbol{X}_i$ 是 $p\times 1$ 解释变量, $\boldsymbol{\beta} = (\beta_1, \beta_2, \cdots, \beta_p)^{\mathrm{T}}$ 是 p- 维未知参数, Z_{ki} 是自变量观测值, $m_k(\cdot)$ 是未知的光滑函数, ε_i 是均值为 0, 方差为 σ^2 的独立随机误差. 为了保证模型的可识别性, 假定 $E\{m_k(Z_k)\} = 0, k = 1, 2, \cdots, q$. 不失一般性, 假定 Y_i 和 $\boldsymbol{X}_i$ 是中心化后的. 显然, 当 $q = 1$ 时, 模型 (7.1) 即为 Engle 等 (1986) 提出的部分线性模型, 关于该模型的详细介绍可参考 Härdle 等 (2000).

由于半参数可加模型的复杂性, 其不但含有参数部分还含有多个非参数函数, 关于该模型的研究结果还相对较少. 相比于部分线性模型以及变系数模型等, 该模型的研究结果还主要集中于模型的估计以及基本的检验问题上. Opsomer 和 Ruppert (1999) 基于 backfitting 技术给出了参数分量的估计, 但是提出的估计量要想达到 $\sqrt{n}$ 相合必须附加 "光滑不足 (under smoothing)" 条件. Li (2000) 基于级数方法研究了模型的估计并证明了估计量的渐近性质. Fan 等 (1998) 以及 Fan 和 Li (2003) 提出利用核方法对模型进行估计. Manzana 和 Zerom (2005) 利用工具变量对模型的参数分量进行估计. 在模型的检验问题上, Härdle 等 (2004) 利用 bootstrap 方法和边际积分 (marginal intergation) 技术研究了广义半参数可加模型的检验问题. Jiang 等 (2007) 基于 Fan, Zhang 和 Zhang (2001) 提出的广义似然比方法和 Opsomer 和 Ruppert (1999) 提出的 backfitting 估计技术研究了模型中非参数分量的检验问题. Deng 和 Liang (2010) 基于多项式样条和模型平均方法 (model averaging method) 研究了该模型, Zhang 和 Liang (2011) 基于多项式样条方法研究了广义半参数可加模型的选择以及模型平均问题. Liu 等 (2011) 利用多项式样条对模型进行估计, 并利用 Fan 和 Li (2001) 提出的 smoothly clipped absolute deviation(SCAD) 惩罚似然方法进行参数分量部分的变量选择. Wang 等 (2011) 利用多项式样条和 SCAD 方法研究了广义半参数可加模型. Ma 和 Yang (2011) 基于样条方法和核方法提出了一种 Spline-backfitted 核光滑估计方法, 所得非参数分量估计具有先知性质, 即其性质和其他非参数分量完全已知时相同. Chen 等 (2011) 基于惩罚样条方法讨论了半参数可加模型的设定检验问题, 构造了一种广义 F 检验统计量.

Liang 等 (2008) 在自变量 $\boldsymbol{X}$ 测量含误差的情形下对参数分量 $\boldsymbol{\beta}$ 提出了校正 profile 最小二乘估计, 该估计是 $\sqrt{n}$ 相合的, 且不需要光滑不足的条件. 相比而言, Opsomer 和 Ruppert (1999) 提出的 backfitting 要达到 $\sqrt{n}$ 相合, 需要非参数分量是欠光滑的. 本章首先将 Liang 等 (2008) 的估计方法推广到一般 $q > 2$ 的情形. 结果表明得到的 profile 最小二乘估计是渐近正态的并且不需要欠光滑的条件.

此外, 实际数据分析中, 能从外部信息得到关于回归系数的一些先验信息, 使用这些信息可以提高估计的有效性, 相关的例子可参考 Rao 和 Toutenburg (1999). 众所周知, 当对一般线性回归模型中的回归系数附加精确的线性约束时, 可以构造比普通的最小二乘估计更有效的约束最小二乘估计, 目前关于约束半参数模型的研究结果还很少. 对于部分线性模型, Przystalski 和 Krajewski (2007) 给出了参数分量的约束估计. 然而约束估计的渐近性质和有效性没有讨论. 本章将研究模型 (7.1) 在含有约束条件时的估计问题. 假定对参数部分附加有如下的约束条件

$$\boldsymbol{A\beta} = \boldsymbol{b}, \tag{7.2}$$

其中 $\boldsymbol{A}$ 是 $k \times p$ 的已知矩阵, 且 $\operatorname{rank}(\boldsymbol{A}) = k$, $\boldsymbol{b}$ 是 $k \times 1$ 已知向量. 将对模型 (7.1) 在附加约束条件 (7.2) 时提出参数分量的约束估计并研究估计量的性质.

实际数据分析中, 经常需要检验约束条件的存在性. 为简单起见, 考虑下面的线性假设

$$H_0 : \boldsymbol{A\beta} = \boldsymbol{0} \quad \text{VS} \quad H_1 : \boldsymbol{A\beta} \neq \boldsymbol{0}. \tag{7.3}$$

这类检验问题的研究也是回归分析的一个重要方面, 很多种具体的检验问题都可以转为上述形式的检验. 根据 $\boldsymbol{\beta}$ 估计量的渐近性质, 可以构造 Wald 型检验统计量. 本章将尝试应用广义似然比 (GLR) 检验这个问题. Fan 等 (2001) 基于多类非参数和半参数模型比如单变量非参数回归模型, 变系数模型提出了 GLR 检验. GLR 检验已经得到了广泛的应用, 包括可加模型 (Fan and Jiang, 2005), 部分线性变系数模型 (Fan and Huang, 2005), 部分线性单指数模型 (Zhang and Huang, 2009),

带有不同光滑变量的变系数模型 (Ip *et al.*, 2007). 值得一提的是 Jiang 等 (2007) 对模型 (7.1) 使用 GLR 方法讨论了非参数分量的设定检验问题, 该文献使用的是 backfitting 估计方法, 也没有讨论参数分量的检验问题.

7.2 参数分量的 profile 最小二乘估计

本节将基于 Liang 等 (2008) 中的估计方法给出模型 (7.1) 的 profile 最小二乘估计.

假定 $\boldsymbol{\beta}$ 已知, 模型 (7.1) 可以写成如下形式

$$Y_i-\boldsymbol{X}_i^{\mathrm{T}}\boldsymbol{\beta}=m_1(Z_{1i})+\cdots+m_q(Z_{qi})+\varepsilon_i,\quad i=1,2,\cdots,n. \tag{7.4}$$

显然模型 (7.4) 是一个标准的可加模型. 对于该模型, 采用 backfitting 方法估计里面的未知非参数函数. 关于 backfitting 方法的详细讨论可参考 Buja 等 (1989), Hastie 和 Tibshirani (1990) 以及 Opsomer (2000).

记

$$\boldsymbol{Y}=\begin{pmatrix}Y_1\\Y_2\\\vdots\\Y_n\end{pmatrix},\quad \boldsymbol{X}=\begin{pmatrix}\boldsymbol{X}_1^{\mathrm{T}}\\\boldsymbol{X}_2^{\mathrm{T}}\\\vdots\\\boldsymbol{X}_n^{\mathrm{T}}\end{pmatrix},\quad \boldsymbol{m}_k=\begin{pmatrix}m_k(Z_{k1})\\m_k(Z_{k2})\\\vdots\\m_k(Z_{kn})\end{pmatrix},\quad \boldsymbol{D}_{Z_k}^k=\begin{pmatrix}1&(Z_{k1}-Z_k)\\1&(Z_{k2}-Z_k)\\\vdots&\vdots\\1&(Z_{kn}-Z_k)\end{pmatrix}.$$

由局部线性方法的原理, 关于第 $k(k=1,2,\cdots,q)$ 个非参数函数的局部线性方法光滑矩阵可记为

$$\boldsymbol{S}_k=\begin{pmatrix}\boldsymbol{e}_1^{\mathrm{T}}\{\boldsymbol{D}_{Z_{k1}}^{k\mathrm{T}}\boldsymbol{K}_{Z_{k1}}\boldsymbol{D}_{Z_{k1}}^k\}^{-1}\boldsymbol{D}_{Z_{k1}}^{\mathrm{T}}\boldsymbol{K}_{Z_{k1}}\\\boldsymbol{e}_1^{\mathrm{T}}\{\boldsymbol{D}_{Z_{k2}}^{k\mathrm{T}}\boldsymbol{K}_{Z_{k2}}\boldsymbol{D}_{Z_{k2}}^k\}^{-1}\boldsymbol{D}_{Z_{k2}}^{\mathrm{T}}\boldsymbol{K}_{Z_{k2}}\\\vdots\\\boldsymbol{e}_1^{\mathrm{T}}\{\boldsymbol{D}_{Z_{kn}}^{k\mathrm{T}}\boldsymbol{K}_{Z_{kn}}\boldsymbol{D}_{Z_{kn}}^k\}^{-1}\boldsymbol{D}_{Z_{kn}}^{\mathrm{T}}\boldsymbol{K}_{Z_{kn}}\end{pmatrix},$$

其中 $\boldsymbol{e}_1=(1,0)^{\mathrm{T}}$, $\boldsymbol{K}_{Z_k}=\mathrm{diag}(K_{h_k}(Z_{k1}-Z_k),K_{h_k}(Z_{k2}-Z_k),\cdots,K_{h_k}(Z_{kn}-Z_k))$, 对于核函数 $K(\cdot)$ 和窗宽 h_k 有, $K_{h_k}(\cdot)=K(\cdot/h_k)/h_k$. 通过解下面的正规方程可以得到未知非参数函数 $\boldsymbol{m}_k$ 的估计

$$\begin{pmatrix}\boldsymbol{I}_n&\boldsymbol{S}_1^*&\cdots&\boldsymbol{S}_1^*\\\boldsymbol{S}_2^*&\boldsymbol{I}_n&\cdots&\boldsymbol{S}_2^*\\\vdots&\vdots&&\vdots\\\boldsymbol{S}_q^*&\boldsymbol{S}_q^*&\cdots&\boldsymbol{I}_n\end{pmatrix}\begin{pmatrix}\boldsymbol{m}_1\\\boldsymbol{m}_2\\\vdots\\\boldsymbol{m}_q\end{pmatrix}=\begin{pmatrix}\boldsymbol{S}_1^*\\\boldsymbol{S}_2^*\\\vdots\\\boldsymbol{S}_q^*\end{pmatrix}(\boldsymbol{Y}-\boldsymbol{X}\boldsymbol{\beta}), \tag{7.5}$$

其中 $\boldsymbol{S}_k^* = (\boldsymbol{I}_n - \boldsymbol{1}\boldsymbol{1}^{\mathrm{T}}/n)\boldsymbol{S}_k$ 是 $\boldsymbol{S}_k$ 相应的中心化光滑矩阵, $\boldsymbol{1}_n$ 是元素全为 1 的 n 维列向量. 如果 $\boldsymbol{S}^{-1}$ 存在, Opsomer (2000) 指出非参数函数的 backfitting 估计可记为

$$\begin{pmatrix} \hat{\boldsymbol{m}}_1 \\ \hat{\boldsymbol{m}}_2 \\ \vdots \\ \hat{\boldsymbol{m}}_q \end{pmatrix} = \begin{pmatrix} \boldsymbol{I}_n & \boldsymbol{S}_1^* & \cdots & \boldsymbol{S}_1^* \\ \boldsymbol{S}_2^* & \boldsymbol{I}_n & \cdots & \boldsymbol{S}_2^* \\ \vdots & \vdots & & \vdots \\ \boldsymbol{S}_q^* & \boldsymbol{S}_q^* & \cdots & \boldsymbol{I}_n \end{pmatrix}^{-1} \begin{pmatrix} \boldsymbol{S}_1^* \\ \boldsymbol{S}_2^* \\ \vdots \\ \boldsymbol{S}_q^* \end{pmatrix} (\boldsymbol{Y} - \boldsymbol{X\beta}) \equiv \boldsymbol{S}^{-1}\boldsymbol{C}(\boldsymbol{Y} - \boldsymbol{X\beta}), \tag{7.6}$$

Opsomer (2000) 中证明了 backfitting 估计的存在性和唯一性.

此外由 Opsomer (2000), $\boldsymbol{m}_k$ 的估计可表示为 $\hat{\boldsymbol{m}}_k = \boldsymbol{W}_k(\boldsymbol{Y} - \boldsymbol{X\beta})$, 其中

$$\boldsymbol{W}_k = \boldsymbol{E}_k\boldsymbol{S}^{-1}\boldsymbol{C}, \quad k = 1, 2, \cdots, q, \tag{7.7}$$

$\boldsymbol{E}_k$ 是一个 $n \times nq$ 维的分块矩阵, 第 k 块是一个 $n \times n$ 的单位矩阵, 其他部分都是 0. 记 $\boldsymbol{M} = \sum_{k=1}^{q} \boldsymbol{m}_k$ 和 $\boldsymbol{W}_{\boldsymbol{M}} = \sum_{k=1}^{q} \boldsymbol{W}_k$, 则 $\boldsymbol{M}$ 的 backfitting 估计为

$$\hat{\boldsymbol{M}} = \boldsymbol{W}_{\boldsymbol{M}}(\boldsymbol{Y} - \boldsymbol{X\beta}). \tag{7.8}$$

用 $\hat{\boldsymbol{M}}$ 代替模型 (7.4) 中 $\boldsymbol{M}$, 可得如下的线性回归模型

$$Y_i - \tilde{Y}_i = (\boldsymbol{X}_i - \tilde{\boldsymbol{X}}_i)^{\mathrm{T}}\boldsymbol{\beta} + \varepsilon_i, \tag{7.9}$$

其中 $\tilde{\boldsymbol{Y}} = (\tilde{Y}_1, \cdots, \tilde{Y}_n)^{\mathrm{T}} = \boldsymbol{W}_{\boldsymbol{M}}\boldsymbol{Y}$, $\tilde{\boldsymbol{X}} = (\tilde{\boldsymbol{X}}_1, \cdots, \tilde{\boldsymbol{X}}_n)^{\mathrm{T}} = \boldsymbol{W}_{\boldsymbol{M}}\boldsymbol{X}$. 基于线性模型 (7.9), 由最小二乘估计方法可得参数分量 $\boldsymbol{\beta}$ 的 profile 最小二乘估计

$$\begin{aligned} \hat{\boldsymbol{\beta}} &= \left\{ \sum_{i=1}^{n} (\boldsymbol{X}_i - \tilde{\boldsymbol{X}}_i)(\boldsymbol{X}_i - \tilde{\boldsymbol{X}}_i)^{\mathrm{T}} \right\}^{-1} \sum_{i=1}^{n} (\boldsymbol{X}_i - \tilde{\boldsymbol{X}}_i)(Y_i - \tilde{Y}_i) \\ &= \left[\boldsymbol{X}^{\mathrm{T}}(\boldsymbol{I}_n - \boldsymbol{W}_{\boldsymbol{M}})^{\mathrm{T}}(\boldsymbol{I}_n - \boldsymbol{W}_{\boldsymbol{M}})\boldsymbol{X} \right]^{-1} \boldsymbol{X}^{\mathrm{T}}(\boldsymbol{I}_n - \boldsymbol{W}_{\boldsymbol{M}})^{\mathrm{T}}(\boldsymbol{I}_n - \boldsymbol{W}_{\boldsymbol{M}})\boldsymbol{Y}. \end{aligned} \tag{7.10}$$

相应地, $\boldsymbol{M}$ 的估计为

$$\hat{\boldsymbol{M}} = \boldsymbol{W}_{\boldsymbol{M}}(\boldsymbol{Y} - \boldsymbol{X}\hat{\boldsymbol{\beta}}). \tag{7.11}$$

定义 $\hat{\boldsymbol{Y}} = (\hat{y}_1, \hat{y}_2, \cdots, \hat{y}_n)^{\mathrm{T}}$ 为因变量 $\boldsymbol{Y}$ 的拟合值, $\hat{\boldsymbol{\varepsilon}} = (\hat{\varepsilon}_1, \hat{\varepsilon}_2, \cdots, \hat{\varepsilon}_n)^{\mathrm{T}}$ 是残差向量. 根据上述估计方法, 以及式 (7.10) 和式 (7.11), 可得

$$\hat{\boldsymbol{Y}} = \boldsymbol{X}\hat{\boldsymbol{\beta}} + \boldsymbol{W}_{\boldsymbol{M}}(\boldsymbol{Y} - \boldsymbol{X}\hat{\boldsymbol{\beta}}) = \boldsymbol{LY}, \quad \hat{\boldsymbol{\varepsilon}} = \boldsymbol{Y} - \hat{\boldsymbol{Y}} = (\boldsymbol{I}_n - \boldsymbol{L})\boldsymbol{Y}, \tag{7.12}$$

其中

$$\boldsymbol{L} = \boldsymbol{W_M} + \overline{\boldsymbol{X}}[\overline{\boldsymbol{X}}^{\mathrm{T}}\overline{\boldsymbol{X}}]^{-1}\overline{\boldsymbol{X}}^{\mathrm{T}}(\boldsymbol{I}_n - \boldsymbol{W_M}). \tag{7.13}$$

和 $\overline{\boldsymbol{X}} = (\boldsymbol{I}_n - \boldsymbol{W_M})\boldsymbol{X} = \boldsymbol{X} - \widetilde{\boldsymbol{X}}$. 上述的估计依赖于窗宽 h_k, 注意到上面的拟合方法为线性拟合, 因此, 根据 Hastie 和 Tibishrani (1990, 3.4.3 节), 可以用交叉证实 (CV) 方法来选择窗宽. 所选定的窗使得下式达到最小

$$\mathrm{CV}(\boldsymbol{h}) = \sum_{i=1}^{n}\left(\frac{\hat{\varepsilon}_i}{1-l_{ii}}\right)^2,$$

其中 $\boldsymbol{h} = (h_1, h_2, \cdots, h_q)^{\mathrm{T}}$, l_{ii} 是矩阵 $\boldsymbol{L}$ 的第 i 个对角元素.

下面的定理给出了 $\hat{\boldsymbol{\beta}}$ 的渐近性质.

定理 7.1　若 7.6 节中的条件 1-4 成立, $\boldsymbol{\beta}$ 的 profile 最小二乘估计是渐近正态的,

$$\sqrt{n}(\hat{\boldsymbol{\beta}} - \boldsymbol{\beta}) \xrightarrow{D} N(\boldsymbol{0}, \sigma^2\boldsymbol{\Sigma}^{-1}),$$

其中 $\boldsymbol{\Sigma} = E\left[\boldsymbol{X}_i - \sum_{k=1}^{q} E(\boldsymbol{X}_i|Z_{ki})\right]\left[\boldsymbol{X}_i - \sum_{k=1}^{q} E(\boldsymbol{X}_i|Z_{ki})\right]^{\mathrm{T}}$.

上面给出了参数分量估计量的性质, 对于非参数分量 $\{m_k(\cdot), k = 1, 2, \cdots, q\}$ 来说, 可以采用两步估计估计得到它们的估计值. 首先通过提出的 profile 最小二乘方法得到 $\boldsymbol{\beta}$ 的估计, 然后用 $\hat{\boldsymbol{\beta}}$ 替换模型 (7.4) 中的 $\boldsymbol{\beta}$, 此时的模型 (7.4) 本质上就是一个将 $Y_i - \boldsymbol{X}_i^{\mathrm{T}}\hat{\boldsymbol{\beta}}$ 作为新的因变量的标准可加模型, 接下来再用 backfitting 方法得到非参数分量的最终估计. 定理 7.1 显示 $\hat{\boldsymbol{\beta}}$ 是 $\boldsymbol{\beta}$ 的 $\sqrt{n}$ 相合估计, 因此这样构造的非参数函数估计量和 Opsomer (2000) 中针对纯可加模型得到的非参数分量估计量的渐近性质是一样的.

注 7.1　基于 backfitting 方法, Opsomer 和 Ruppert (1999) 对参数分量 $\boldsymbol{\beta}$ 提出了一种 backfitting 估计

$$\hat{\boldsymbol{\beta}}_{\mathrm{bf}} = \left[\boldsymbol{X}^{\mathrm{T}}(\boldsymbol{I}_n - \boldsymbol{W_M})\boldsymbol{X}\right]^{-1}\boldsymbol{X}^{\mathrm{T}}(\boldsymbol{I}_n - \boldsymbol{W_M})\boldsymbol{Y}. \tag{7.14}$$

Opsomer 和 Ruppert (1999) 证明了在通常的窗宽 $h \sim n^{-1/5}$ 下, backfitting 估计 $\hat{\boldsymbol{\beta}}_{\mathrm{bf}}$ 不是 $\sqrt{n}$ 相合的, 所以必须附加光滑不足条件. 不同于 $\hat{\boldsymbol{\beta}}_{\mathrm{bf}}$, 我们所提出的 profile 最小二乘估计 $\hat{\boldsymbol{\beta}}$ 在通常 $h \sim n^{-1/5}$ 下是 $\sqrt{n}$ 相合的.

7.3　参数分量的约束 profile 最小二乘估计

在上述构造 profile 最小二乘估计过程中, 没有用到约束条件 $\boldsymbol{A\beta} = \boldsymbol{b}$, 这可能会导致信息的丢失. 为了解决这个问题, 下面将构造一种相应的约束估计, 它不仅

是相合的而且满足约束条件 (7.2). 基于线性 (7.9), 利用 Lagrange 乘子方法, 定义相对应于约束条件 $\boldsymbol{A\beta}=\boldsymbol{b}$ 的 Lagrange 函数

$$F(\boldsymbol{\beta},\boldsymbol{\lambda})=(\overline{\boldsymbol{Y}}-\overline{\boldsymbol{X}}\boldsymbol{\beta})^{\mathrm{T}}(\overline{\boldsymbol{Y}}-\overline{\boldsymbol{X}}\boldsymbol{\beta})+2\boldsymbol{\lambda}^{\mathrm{T}}(\boldsymbol{A\beta}-\boldsymbol{b}), \tag{7.15}$$

其中 $\overline{\boldsymbol{Y}}=\boldsymbol{Y}-\tilde{\boldsymbol{Y}}$, $\boldsymbol{\lambda}$ 是 k 维 Lagrange 乘子. 针对函数 $\boldsymbol{F}(\boldsymbol{\beta},\boldsymbol{\lambda})$ 关于 $\boldsymbol{\beta}$ 和 $\boldsymbol{\lambda}$ 分别求偏导, 并令偏导数等于 $\mathbf{0}$, 有

$$\begin{cases} \dfrac{\partial F(\boldsymbol{\beta},\boldsymbol{\lambda})}{\partial\boldsymbol{\beta}}=-2\overline{\boldsymbol{X}}^{\mathrm{T}}\overline{\boldsymbol{Y}}+2\overline{\boldsymbol{X}}^{\mathrm{T}}\overline{\boldsymbol{X}}\boldsymbol{\beta}+2\boldsymbol{A}^{\mathrm{T}}\boldsymbol{\lambda}=\mathbf{0}, \\ \dfrac{\partial F(\boldsymbol{\beta},\boldsymbol{\lambda})}{\partial\boldsymbol{\lambda}}=2(\boldsymbol{A\beta}-\boldsymbol{b})=\mathbf{0}. \end{cases} \tag{7.16}$$

求解方程组 (7.16), 得 $\boldsymbol{\beta}$ 的约束 profile 最小二乘估计为

$$\hat{\boldsymbol{\beta}}_r=\hat{\boldsymbol{\beta}}-(\overline{\boldsymbol{X}}^{\mathrm{T}}\overline{\boldsymbol{X}})^{-1}\boldsymbol{A}^{\mathrm{T}}\left[\boldsymbol{A}(\overline{\boldsymbol{X}}^{\mathrm{T}}\overline{\boldsymbol{X}})^{-1}\boldsymbol{A}^{\mathrm{T}}\right]^{-1}(\boldsymbol{A}\hat{\boldsymbol{\beta}}-\boldsymbol{b}). \tag{7.17}$$

相应地, $\boldsymbol{M}$ 的约束估计为

$$\hat{\boldsymbol{M}}_r=\boldsymbol{W}_{\boldsymbol{M}}(\boldsymbol{Y}-\boldsymbol{X}\hat{\boldsymbol{\beta}}_r). \tag{7.18}$$

下面给出 $\hat{\boldsymbol{\beta}}_r$ 的渐近性质.

定理 7.2 若 7.6 中的条件 1~ 条件 4 成立, $\boldsymbol{\beta}$ 的约束 profile 最小二乘估计是渐近正态的,

$$\sqrt{n}(\hat{\boldsymbol{\beta}}_r-\boldsymbol{\beta})\xrightarrow{d}N(\mathbf{0},\sigma^2\boldsymbol{\Omega}),$$

其中 $\boldsymbol{\Omega}=\boldsymbol{\Sigma}^{-1}-\boldsymbol{\Sigma}^{-1}\boldsymbol{A}^{\mathrm{T}}\left(\boldsymbol{A}\boldsymbol{\Sigma}^{-1}\boldsymbol{A}^{\mathrm{T}}\right)^{-1}\boldsymbol{A}\boldsymbol{\Sigma}^{-1}$.

由于 $\boldsymbol{\Delta}=\boldsymbol{\Sigma}^{-1}\boldsymbol{A}^{\mathrm{T}}\left(\boldsymbol{A}\boldsymbol{\Sigma}^{-1}\boldsymbol{A}^{\mathrm{T}}\right)^{-1}\boldsymbol{A}\boldsymbol{\Sigma}^{-1}$ 半正定矩阵. 因此若约束条件 (7.2) 成立, $\hat{\boldsymbol{\beta}}_r$ 比 $\hat{\boldsymbol{\beta}}$ 更有效.

注 7.2 基于 Opsomer 和 Ruppert (1999) 中的 backfitting 估计方法, 也可以对参数分量 $\boldsymbol{\beta}$ 构造约束 backfitting 估计. 首先 $\boldsymbol{\beta}$ 的 backfitting 估计 $\hat{\boldsymbol{\beta}}_{\mathrm{bf}}$ 满足

$$\hat{\boldsymbol{\beta}}_{\mathrm{bf}}=\arg\min_{\boldsymbol{\beta}\in\mathbf{R}^p}\left[(\boldsymbol{Y}-\boldsymbol{X\beta})^{\mathrm{T}}(\boldsymbol{I}_n-\boldsymbol{W}_{\boldsymbol{M}})(\boldsymbol{Y}-\boldsymbol{X\beta})\right].$$

为了构造参数分量的约束 backfitting 估计, 定义 Lagrange 函数

$$\overline{F}(\boldsymbol{\beta},\boldsymbol{\lambda})=(\boldsymbol{Y}-\boldsymbol{X\beta})^{\mathrm{T}}(\boldsymbol{I}_n-\boldsymbol{W}_{\boldsymbol{M}})(\boldsymbol{Y}-\boldsymbol{X\beta})+2\boldsymbol{\lambda}^{\mathrm{T}}(\boldsymbol{A\beta}-\boldsymbol{b}).$$

将 $\overline{F}(\boldsymbol{\beta},\boldsymbol{\lambda})$ 分别对 $\boldsymbol{\beta}$ 和 $\boldsymbol{\lambda}$ 求偏导, 并使票导数等于 $\mathbf{0}$, 可得

$$\begin{cases} \dfrac{\partial\overline{F}(\boldsymbol{\beta},\boldsymbol{\lambda})}{\partial\boldsymbol{\beta}}=-2\boldsymbol{X}^{\mathrm{T}}(\boldsymbol{I}_n-\boldsymbol{W}_{\boldsymbol{M}})(\boldsymbol{Y}-\boldsymbol{X\beta})+2\boldsymbol{A}^{\mathrm{T}}\boldsymbol{\lambda}=\mathbf{0}, \\ \dfrac{\partial F(\boldsymbol{\beta},\boldsymbol{\lambda})}{\partial\boldsymbol{\lambda}}=2(\boldsymbol{A\beta}-\boldsymbol{b})=\mathbf{0}. \end{cases}$$

求解上面的方程组, 得 $\boldsymbol{\beta}$ 的约束 backfitting 估计为

$$\hat{\boldsymbol{\beta}}_{\mathrm{bf}}^{r}=\hat{\boldsymbol{\beta}}_{\mathrm{bf}}-\left[\boldsymbol{X}^{\mathrm{T}}(\boldsymbol{I}-\boldsymbol{W}_{\boldsymbol{M}})\boldsymbol{X}\right]^{-1}\boldsymbol{A}^{\mathrm{T}}\left(\boldsymbol{A}[\boldsymbol{X}^{\mathrm{T}}(\boldsymbol{I}-\boldsymbol{W}_{\boldsymbol{M}})\boldsymbol{X}]^{-1}\boldsymbol{A}^{\mathrm{T}}\right)^{-1}\left(\boldsymbol{A}\hat{\boldsymbol{\beta}}_{\mathrm{bf}}-\boldsymbol{b}\right).$$

7.4 广义似然比检验

针对假设检验问题 (7.3), 本节将构造 GLR 检验统计量, 并研究它的性质.

首先假定 $\boldsymbol{\varepsilon}\sim N(\mathbf{0},\sigma^2\boldsymbol{I}_n)$. 正如 Fan 等 (2001) 和 Jiang 等 (2007) 所说, 正态性假设仅仅是为了构造统计量的需要, 统计量性质的证明中并不依赖这一假设. 备择假设下模型 (7.1) 的对数似然函数为

$$L(H_1)=-\frac{n}{2}-\frac{n}{2}\log(2\pi/n)-\frac{n}{2}\log\mathrm{RSS}(\mathrm{H}_1),\tag{7.19}$$

其中 $\mathrm{RSS}(H_1)=\|\boldsymbol{Y}-\boldsymbol{X}\hat{\boldsymbol{\beta}}-\hat{\boldsymbol{M}}\|^2$. 此外, 如果原假设 H_0 成立, 通过 7.3 节中的约束修正 profile 最小二乘方法可得 $\boldsymbol{\beta}$ 和 $\boldsymbol{M}$ 的估计为

$$\hat{\boldsymbol{\beta}}_0=\hat{\boldsymbol{\beta}}-(\overline{\boldsymbol{X}}^{\mathrm{T}}\overline{\boldsymbol{X}})^{-1}\boldsymbol{A}^{\mathrm{T}}\left[\boldsymbol{A}(\overline{\boldsymbol{X}}^{\mathrm{T}}\overline{\boldsymbol{X}})^{-1}\boldsymbol{A}^{\mathrm{T}}\right]^{-1}\boldsymbol{A}\hat{\boldsymbol{\beta}},\quad \hat{\boldsymbol{M}}_0=\boldsymbol{W}_{\boldsymbol{M}}(\boldsymbol{Y}-\boldsymbol{X}\hat{\boldsymbol{\beta}}_0),\tag{7.20}$$

因此, 原假设下模型 (7.1) 的对数似然函数是

$$L(H_0)=-\frac{n}{2}-\frac{n}{2}\log(2\pi/n)-\frac{n}{2}\log\mathrm{RSS}(\mathrm{H}_0),\tag{7.21}$$

其中 $\mathrm{RSS}(H_0)=\|\boldsymbol{Y}-\boldsymbol{X}\hat{\boldsymbol{\beta}}_0-\hat{\boldsymbol{M}}_0\|^2$.

沿用 Fan 等 (2001) 和 Jiang 等 (2007) 中的方法, 定义如下的 profile 广义似然比统计量

$$T_n=L(H_1)-L(H_0)=\frac{n}{2}\log\frac{\mathrm{RSS}(H_0)}{\mathrm{RSS}(H_1)}\approx\frac{n}{2}\frac{\mathrm{RSS}(H_0)-\mathrm{RSS}(H_1)}{\mathrm{RSS}(H_1)}.\tag{7.22}$$

显然, 如果 H_0 成立, $\mathrm{RSS}(H_0)$ 和 $\mathrm{RSS}(H_1)$ 之间不应该有显著的差距. 下面的两个定理分别给出了检验统计量 T_n 的渐近零分布及其功效.

定理 7.3 若 7.6 节中的条件 1~条件 4 成立, 检验问题 (7.3) 中的原假设若成立, 则有

$$2T_n\xrightarrow{D}\chi_k^2,$$

其中 χ_k^2 自由度为 k 的 χ^2 分布.

定理 7.4 若 7.6 节中的条件 1~ 条件 4 成立, 在检验问题 (7.3) 中备择假设下有

$$2T_n\xrightarrow{D}\chi_k^2(\lambda),$$

其中 $\chi_k^2(\lambda)$ 是一个自由度为 k 非中心 χ^2 分布, 非中心参数 $\lambda = \lim\limits_{n\to\infty} n\sigma^{-2}\boldsymbol{\beta}^{\mathrm{T}}\boldsymbol{A}^{\mathrm{T}}\left(\boldsymbol{A}\boldsymbol{\Sigma}^{-1}\boldsymbol{A}^{\mathrm{T}}\right)^{-1}\boldsymbol{A}\boldsymbol{\beta}$.

注 7.3 对于假设检验问题 (7.3), 可构造 Wald 类型的检验统计量

$$W_n = n\hat{\boldsymbol{\beta}}^{\mathrm{T}}\boldsymbol{A}^{\mathrm{T}}\left(\boldsymbol{A}\hat{\boldsymbol{\Xi}}\boldsymbol{A}^{\mathrm{T}}\right)^{-1}\boldsymbol{A}\hat{\boldsymbol{\beta}}.$$

可以证明 W_n 和 $2T_n$ 有相同的渐近分布.

7.5 数值模拟

本节将通过数值模拟来考察所提方法的有效性. 假设数据来自于如下的部分线性可加模型

$$y_i = x_{1i}\beta_1 + x_{2i}\beta_2 + m_1(z_{1i}) + m_2(z_{2i}) + \varepsilon_i, \quad i = 1, 2, \cdots, n, \tag{7.23}$$

其中 $m_1(z_{1i}) = \sin(2\pi z_{1i})$, $m_2(z_{2i}) = z_{2i}^3 + 3z_{2i}^2 - 5z_{2i} - 1$, $x_{1i} \sim N(0,1)$, $x_{2i} \sim U(-2,2)$, $z_{1i} \sim U(0,1)$, $z_{2i} \sim U(-1,1)$. 为了考察误差分布的影响, 采取以下三种不同类型的误差分布, 它们的方差都是 $\sigma^2 = 0.25$, ① $\varepsilon_i \sim N(0, 0.5^2)$, ② $\varepsilon_i \sim U(-\sqrt{3}/2, \sqrt{3}/2)$, ③ $\varepsilon_i \sim \dfrac{\sqrt{3}}{4}t(8)$. 模拟中采用 Epanechnikov 核函数: $K(x) = 0.75(1-x^2)\boldsymbol{I}_{|x|\leqslant 1}$. 窗宽的选择采用 7.2 节中的 CV 方法.

无约束估计和有约束估计的比较 为了比较约束估计和没考虑约束条件的无约束估计, 令 $\beta_1 = 3, \beta_2 = 2$, 设定约束条件为 $\beta_1 + \beta_2 = 5$. 利用前面介绍的估计方法分别得到 profile 最小二乘估计 $\hat{\boldsymbol{\beta}}$ 和相应的约束 profile 最小二乘估计 $\hat{\boldsymbol{\beta}}_r$, backfitting 估计 $\hat{\boldsymbol{\beta}}_{\mathrm{bf}}$ 和相应的约束 backfitting 估计 $\hat{\boldsymbol{\beta}}_{\mathrm{bf}}^r$. 设定样本容量为 $n = 150, 200$, 对每一种情况, 重复 1000 次. 模拟结果见表 7.1 和表 7.2, 其中 Mean 代表 β 1000 次估计的平均值, SD 表示 β 1000 次估计值的标准差. 不难看出参数的所有估计值都接近真实值, 随着样本量的增加, 所有估计的偏差和 SD 都在减小. 另外 $\hat{\boldsymbol{\beta}}_r$ 和 $\hat{\boldsymbol{\beta}}_{\mathrm{bf}}^r$ 分

表 7.1 无约束和约束 profile 最小二乘估计的比较

β	误差	n	$\hat{\boldsymbol{\beta}}$			$\hat{\boldsymbol{\beta}}^r$		
			均值	SD	MSE	均值	SD	MSE
$\beta_1 = 3$	$N(0, 0.5^2)$	150	2.998	0.050	0.002	3.000	0.032	0.001
		200	3.001	0.041	0.001	3.000	0.026	0.000
	$U(-\frac{\sqrt{3}}{2}, \frac{\sqrt{3}}{2})$	150	3.000	0.048	0.002	2.998	0.030	0.000
		200	3.001	0.038	0.001	3.000	0.025	0.000
	$\frac{\sqrt{3}}{4}t(8)$	150	2.997	0.048	0.002	2.998	0.032	0.001
		200	3.001	0.037	0.001	3.000	0.025	0.000

续表

β	误差	n	$\hat{\boldsymbol{\beta}}$			$\hat{\boldsymbol{\beta}}^r$		
			均值	SD	MSE	均值	SD	MSE
$\beta_2=2$	$N(0,0.5^2)$	150	1.998	0.042	0.001	1.999	0.032	0.001
		200	2.000	0.035	0.001	1.999	0.026	0.000
	$U(-\frac{\sqrt{3}}{2},\frac{\sqrt{3}}{2})$	150	2.002	0.042	0.001	2.001	0.030	0.000
		200	2.000	0.035	0.001	1.999	0.025	0.000
	$\frac{\sqrt{3}}{4}t(8)$	150	2.002	0.043	0.001	2.001	0.032	0.001
		200	1.999	0.034	0.001	1.999	0.025	0.000

表 7.2　无约束和约束 backfitting 估计的比较

β	误差	n	$\hat{\boldsymbol{\beta}}_{\mathrm{bf}}$			$\hat{\boldsymbol{\beta}}^r_{\mathrm{bf}}$		
			均值	SD	MSE	均值	SD	MSE
$\beta_1=3$	$N(0,0.5^2)$	150	3.000	0.046	0.002	2.998	0.030	0.000
		200	3.001	0.038	0.001	2.999	0.025	0.000
	$U(-\frac{\sqrt{3}}{2},\frac{\sqrt{3}}{2})$	150	3.000	0.046	0.002	2.999	0.030	0.000
		200	3.000	0.038	0.001	3.000	0.027	0.000
	$\frac{\sqrt{3}}{4}t(8)$	150	2.998	0.049	0.002	2.999	0.030	0.000
		200	2.999	0.040	0.001	2.999	0.026	0.000
$\beta_2=2$	$N(0,0.5^2)$	150	2.001	0.040	0.001	2.001	0.030	0.000
		200	2.001	0.035	0.001	2.000	0.025	0.000
	$U(-\frac{\sqrt{3}}{2},\frac{\sqrt{3}}{2})$	150	2.000	0.040	0.001	2.000	0.030	0.000
		200	2.000	0.036	0.001	1.999	0.027	0.000
	$\frac{\sqrt{3}}{4}t(8)$	150	2.000	0.041	0.001	2.000	0.030	0.000
		200	2.000	0.034	0.001	2.000	0.026	0.000

别优于 $\hat{\boldsymbol{\beta}}$ 和 $\hat{\boldsymbol{\beta}}_{\mathrm{bf}}$. 此外以发现通过 CV 方法选择窗宽得到的 $\hat{\boldsymbol{\beta}}$ 和 $\hat{\boldsymbol{\beta}}_{\mathrm{bf}}$ 之间没有显著的差异.

profile 最小二乘与估计backfitting 估计的比较　Opsomer 和 Ruppert (1999) 证明在窗宽为 $n^{-r}, -1<r<-1/4$ 时, backfitting 估计 $\hat{\boldsymbol{\beta}}_{\mathrm{bf}}$ 是 $\sqrt{n}$ 相合的. 定理 7.1 表明所提出的 profile 最小二乘估计 $\hat{\boldsymbol{\beta}}$ 在 $h\sim n^{-1/5}$ 下是 $\sqrt{n}$ 相合的. 下面进行一个小的模拟研究来比较固定窗宽下的 profile 最小二乘估计和 backfitting 估计. 考虑如下三种窗宽 $h_1=h_2=n^{-0.2}, n^{-0.3}, n^{-0.4}$. 在此只考虑误差分布为正态分布的情形, 令 $\beta_1=1, \beta_2=0$, 样本量为 $n=100,150,200$, 1000 次重复实验中 β_1 估计值的平均偏差和样本标准偏差见表 7.3. 从结果可以看出, 首先两类估计对窗宽的选择都不是非常敏感. 当 $h_1=h_2=n^{-0.2}$ 时, profile 最小二乘估计明显优于 backfitting 估计.

表 7.3 固定窗宽下 $\hat{\beta}$ 和 $\hat{\beta}_{\rm bf}$ 的经验偏差 (标准差)

样本量	估计	窗宽		
		$n^{-0.2}$	$n^{-0.3}$	$n^{-0.4}$
$n=100$	$\hat{\beta}$	0.0000(0.060)	−0.0006(0.063)	0.0033(0.066)
	$\hat{\beta}_{\rm bf}$	−0.0004(0.064)	−0.0013(0.064)	0.0031(0.065)
$n=150$	$\hat{\beta}$	0.0014(0.048)	−0.0009(0.046)	0.0003(0.049)
	$\hat{\beta}_{\rm bf}$	0.0011(0.050)	−0.0007(0.047)	0.0002(0.048)
$n=200$	$\hat{\beta}$	0.0000(0.040)	0.0005(0.039)	−0.0004(0.041)
	$\hat{\beta}_{\rm bf}$	0.0000(0.042)	0.0009(0.039)	−0.0004(0.041)

GLR 检验方法的功效 对模型 (7.23) 考虑如下的检验问题

$$H_0:\beta_1-\beta_2=0 \quad \text{VS} \quad H_1:\beta_1-\beta_2=c,$$

其中 $c=0,\pm0.1,\pm0.2,\pm0.3$, $\beta_1=1,\beta_2=1-c$. 对每一个给定 c 值和每一种误差分布, 基于样本容量 $n=200$ 重复进行 1000 次. 在显著性水平 $\alpha=0.05$, 计算所提出 GLR 检验和 Wald 检验拒绝原假设的频率, 结果见表 7.4. 从模拟结果不难看出当零假设为真 (也就是 $c=0$), 提出的 GLR 检验和 Wald 建立拒绝原假设的频率接近显著性水平 0.05. 当原假设不成立时, 随着越来越背离原假设, 拒绝 H_0 的频率越来越接近 1. 此外, GLR 方法和 Wald 方法表现接近.

表 7.4 1000 次重复下检验 $H_0:\beta_1-\beta_2=0$ 的拒绝频率

c 值	检验方法	误差		分布
		$N(0,0.5^2)$	$U(-\sqrt{3}/2,\sqrt{3}/2)$	$\frac{\sqrt{3}}{4}t(8)$
0.0	GLR	0.048	0.072	0.066
	Wald	0.056	0.072	0.066
0.1	GLR	0.500	0.508	0.502
	Wald	0.502	0.518	0.504
−0.1	GLR	0.472	0.484	0.488
	Wald	0.476	0.488	0.492
0.2	GLR	0.966	0.942	0.972
	Wald	0.968	0.942	0.972
−0.2	GLR	0.978	0.970	0.956
	Wald	0.980	0.972	0.958
0.3	GLR	1.000	0.998	1.000
	Wald	1.000	0.998	1.000
−0.3	GLR	0.998	1.000	1.000
	Wald	0.998	1.000	1.000

7.6 定理的证明

在给出定理的证明之前, 先给出一些条件.

条件 1 核函数 $K(\cdot)$ 为对称密度函数, 具有紧支撑.

条件 2 Z_k 的密度函数 $f_k(Z_k)$ 是 Lipschitz 连续且有有界支撑 $\boldsymbol{\Omega}_k, k=1,2,\cdots,q$.

条件 3 $m_k(\cdot), k=1,2,\cdots,q$ 具有二阶连续导数的.

条件 4 $n\to\infty$ 时, $h_k\to 0$ 时, $nh_k/\log n\to\infty$ 和 $nh_k^8\to 0, k=1,2,\cdots,q$.

引理 7.1 令 $(\boldsymbol{X}_1,\boldsymbol{Y}_1),\cdots,(\boldsymbol{X}_n,\boldsymbol{Y}_n)$ 为独立同分布 (iid) 的随机序列, 其中 $\boldsymbol{Y}_i, i=1,2,\cdots,n$ 为一元随机变量, 进一步假定 $E|y|^s<\infty$ 与 $\sup_x\int|y|^s f(x,y)dy<\infty$, 其中 f 表示 $(\boldsymbol{X},\boldsymbol{Y})$ 的联合密度. 令 K 为一有界正函数, 并有有界支撑且满足 Lipschitz 条件, 则有

$$\sup_x\left|\frac{1}{n}\sum_{i=1}^n\left[K_h(\boldsymbol{X}_i-x)\boldsymbol{Y}_i-E\left(K_h(\boldsymbol{X}_i-x)\boldsymbol{Y}_i\right)\right]\right|=O_p\left(\left\{\frac{\log(1/h)}{nh}\right\}^{1/2}\right),\tag{7.24}$$

其中对于 $\varepsilon<1-s^{-1}$, 有 $n^{2\varepsilon-1}h\to\infty$.

该引理可由 Mack 和 Silverman (1982) 的结论得到.

引理 7.2 若条件 1-4 成立, 有

$$\frac{1}{n}\sum_{i=1}^n(\boldsymbol{X}_i-\tilde{\boldsymbol{X}}_i)(\boldsymbol{X}_i-\tilde{\boldsymbol{X}}_i)^{\mathrm{T}}\xrightarrow{p}\boldsymbol{\Sigma}.$$

证明 根据 Opsomer (2000) 中的引理 2.1 可得

$$\boldsymbol{W}_{\boldsymbol{M}}=\sum_{k=1}^q\boldsymbol{W}_k=\sum_{k=1}^q\left\{\boldsymbol{I}_n-(\boldsymbol{I}_n-\boldsymbol{S}_k^*\boldsymbol{W}_{\boldsymbol{M}}^{[-k]})^{-1}(\boldsymbol{I}_n-\boldsymbol{S}_k^*)\right\},\tag{7.25}$$

其中 $\boldsymbol{W}_{\boldsymbol{M}}^{[-k]}=\sum_{j\neq k}\boldsymbol{W}_j$. 根据 Opsomer 和 Ruppert(1997) 中的引理 3.1 和引理 3.2, 可得

$$\boldsymbol{S}_k^*=\boldsymbol{S}_k-\mathbf{1}_n\mathbf{1}_n^{\mathrm{T}}/n+o_p(\mathbf{1}_n\mathbf{1}_n^{\mathrm{T}}/n),\quad(\boldsymbol{I}_n-\boldsymbol{S}_k^*\boldsymbol{W}_{\boldsymbol{M}}^{[-k]})^{-1}=\boldsymbol{I}_n+Op(\mathbf{1}_n\mathbf{1}_n^{\mathrm{T}}/n).\tag{7.26}$$

由式 (7.25) 和式 (7.26), 可得

$$\boldsymbol{W}_{\boldsymbol{M}}=\sum_{k=1}^q\boldsymbol{S}_k+Op(\mathbf{1}_n\mathbf{1}_n^{\mathrm{T}}/n).\tag{7.27}$$

由引理 7.1 可证

$$\begin{aligned}\frac{1}{n}\sum_{i=1}^n(\boldsymbol{X}_i-\tilde{\boldsymbol{X}}_i)(\boldsymbol{X}_i-\tilde{\boldsymbol{X}}_i)^{\mathrm{T}}=&\frac{1}{n}\sum_{i=1}^n\left[\boldsymbol{X}_i-\sum_{k=1}^qE(\boldsymbol{X}_i|Z_{ki})\right]\left[\boldsymbol{X}_i-\sum_{k=1}^qE(\boldsymbol{X}_i|Z_{ki})\right]^{\mathrm{T}}\\&+o_p(1)\xrightarrow{p}\boldsymbol{\Sigma}.\end{aligned}$$

引理 7.3 若条件 1~ 条件 4 成立, 有

$$\frac{1}{\sqrt{n}}\boldsymbol{X}^{\mathrm{T}}(\boldsymbol{I}-\boldsymbol{W}_{\boldsymbol{M}})^{\mathrm{T}}(\boldsymbol{I}-\boldsymbol{W}_{\boldsymbol{M}})(\boldsymbol{M}+\boldsymbol{\varepsilon}) \xrightarrow{D} N(\mathbf{0},\sigma^2\boldsymbol{\Sigma}).$$

证明 根据 Fan 和 Jiang (2005) 中的引理 B.6, 有

$$(\boldsymbol{I}_n-\boldsymbol{W}_{\boldsymbol{M}})\boldsymbol{M}=O\left(\sum_{i=1}^{q}h_k^2\mathbf{1}_n+\mathbf{1}_n/\sqrt{n}\right),\quad \text{a.s.,}\quad (\boldsymbol{I}_n-\boldsymbol{W}_{\boldsymbol{M}})\boldsymbol{\varepsilon}$$
$$=\boldsymbol{\varepsilon}-\sum_{i=1}^{q}\boldsymbol{S}_k\boldsymbol{\varepsilon}+O_p\left(n\sum_{i=1}^{q}h_k^4\right).$$

再结合引理 7.2, 很容易证明

$$\frac{1}{n}\boldsymbol{X}^{\mathrm{T}}(\boldsymbol{I}-\boldsymbol{W}_{\boldsymbol{M}})^{\mathrm{T}}(\boldsymbol{I}-\boldsymbol{W}_{\boldsymbol{M}})\boldsymbol{M}=O_p\left(n^{-1/2}\sum_{i=1}^{q}h_k^2\right), \tag{7.28}$$

$$\frac{1}{n}\boldsymbol{X}^{\mathrm{T}}(\boldsymbol{I}-\boldsymbol{W}_{\boldsymbol{M}})^{\mathrm{T}}(\boldsymbol{I}-\boldsymbol{W}_{\boldsymbol{M}})\boldsymbol{\varepsilon}=\frac{1}{n}\sum_{i=1}^{n}\left(\boldsymbol{X}_i-\sum_{k=1}^{q}E(\boldsymbol{X}_i|Z_{ki})\right)\varepsilon_i+O_p\left(n^{-1/2}\sum_{i=1}^{q}h_k^2\right). \tag{7.29}$$

由式 (7.28) 和式 (7.29), 可得

$$\frac{1}{\sqrt{n}}\boldsymbol{X}^{\mathrm{T}}(\boldsymbol{I}-\boldsymbol{W}_{\boldsymbol{M}})^{\mathrm{T}}(\boldsymbol{I}-\boldsymbol{W}_{\boldsymbol{M}})(\boldsymbol{M}+\boldsymbol{\varepsilon})=\frac{1}{\sqrt{n}}\sum_{i=1}^{n}\left(\boldsymbol{X}_i-\sum_{k=1}^{q}E(\boldsymbol{X}_i|Z_{ki})\right)+o_p(1). \tag{7.30}$$

由中心极限定理, 引理 7.3 得证.

定理 7.1 和 7.2 的证明 根据引理 7.2 和引理 7.3, 可直接证明定理 7.1 和定理 7.2, 在此省略.

定理 7.3 和 7.4 的证明 定理 7.3 是定理 7.4 的特例, 因此下面只证定理 7.4. 由 $\mathrm{RSS}(H_1)$ 的定义, 可得

$$\begin{aligned}\frac{1}{n}\mathrm{RSS}(H_1)&=\frac{1}{n}\left[\boldsymbol{Y}-\boldsymbol{X}\hat{\boldsymbol{\beta}}-\hat{\boldsymbol{M}}\right]^{\mathrm{T}}\left[\boldsymbol{Y}-\boldsymbol{X}\hat{\boldsymbol{\beta}}-\hat{\boldsymbol{M}}\right]\\&=\frac{1}{n}\left[\boldsymbol{X}(\boldsymbol{\beta}-\hat{\boldsymbol{\beta}})+\boldsymbol{M}+\boldsymbol{\varepsilon}\right]^{\mathrm{T}}(\boldsymbol{I}-\boldsymbol{W}_{\boldsymbol{M}})^{\mathrm{T}}(\boldsymbol{I}-\boldsymbol{W}_{\boldsymbol{M}})\left[\boldsymbol{X}(\boldsymbol{\beta}-\hat{\boldsymbol{\beta}})+\boldsymbol{M}+\boldsymbol{\varepsilon}\right]\\&=I_1+I_2+I_3+I_4+I_5+I_6,\end{aligned}$$

其中

$$I_1=\frac{1}{n}\boldsymbol{\varepsilon}^{\mathrm{T}}(\boldsymbol{I}-\boldsymbol{W}_{\boldsymbol{M}})^{\mathrm{T}}(\boldsymbol{I}-\boldsymbol{W}_{\boldsymbol{M}})\boldsymbol{\varepsilon},\quad I_2=\frac{1}{n}(\boldsymbol{\beta}-\hat{\boldsymbol{\beta}})^{\mathrm{T}}\overline{\boldsymbol{X}}^{\mathrm{T}}\overline{\boldsymbol{X}}(\boldsymbol{\beta}-\hat{\boldsymbol{\beta}}),$$
$$I_3=\frac{1}{n}\boldsymbol{M}^{\mathrm{T}}(\boldsymbol{I}-\boldsymbol{W}_{\boldsymbol{M}})^{\mathrm{T}}(\boldsymbol{I}-\boldsymbol{W}_{\boldsymbol{M}})\boldsymbol{M},\quad I_4=\frac{2}{n}\boldsymbol{\varepsilon}^{\mathrm{T}}(\boldsymbol{I}-\boldsymbol{W}_{\boldsymbol{M}})^{\mathrm{T}}(\boldsymbol{I}-\boldsymbol{W}_{\boldsymbol{M}})\boldsymbol{M},$$

$$I_5 = \frac{2}{n}(\boldsymbol{\beta} - \hat{\boldsymbol{\beta}})^{\mathrm{T}}\overline{\boldsymbol{X}}^{\mathrm{T}}(\boldsymbol{I} - \boldsymbol{W}_{\boldsymbol{M}})\boldsymbol{M}, \quad I_6 = \frac{2}{n}(\boldsymbol{\beta} - \hat{\boldsymbol{\beta}})^{\mathrm{T}}\overline{\boldsymbol{X}}^{\mathrm{T}}(\boldsymbol{I} - \boldsymbol{W}_{\boldsymbol{M}})\boldsymbol{\varepsilon}.$$

类似于引理 7.2 和引理 7.3 的证明, 易证

$$I_1 = \sigma^2\{1 + o_p(1)\}, \quad I_i = o_p(1), \quad i = 2, 3, \cdots, 6.$$

因此可得

$$\frac{1}{n}\mathrm{RSS}(H_1) = \sigma^2 + o_p(1).$$

由 $\mathrm{RSS}(H_0)$ 的定义, 有

$$\begin{aligned}
\mathrm{RSS}(H_0) &= \left[\boldsymbol{Y} - \boldsymbol{X}\hat{\boldsymbol{\beta}}_0 - \hat{\boldsymbol{M}}_0\right]^{\mathrm{T}}\left[\boldsymbol{Y} - \boldsymbol{X}\hat{\boldsymbol{\beta}}_0 - \hat{\boldsymbol{M}}_0\right] \\
&= \left[\boldsymbol{Y} - \boldsymbol{X}\hat{\boldsymbol{\beta}}_0 - \boldsymbol{W}_{\boldsymbol{M}}(\boldsymbol{Y} - \boldsymbol{X}\hat{\boldsymbol{\beta}}_0)\right]^{\mathrm{T}}\left[\boldsymbol{Y} - \boldsymbol{X}\hat{\boldsymbol{\beta}}_0 - \boldsymbol{W}_{\boldsymbol{M}}(\boldsymbol{Y} - \boldsymbol{X}\hat{\boldsymbol{\beta}}_0)\right] \\
&= \left[\overline{\boldsymbol{Y}} - \overline{\boldsymbol{X}}\hat{\boldsymbol{\beta}}_0\right]^{\mathrm{T}}\left[\overline{\boldsymbol{Y}} - \overline{\boldsymbol{X}}\hat{\boldsymbol{\beta}}_0\right] \\
&= \left[\overline{\boldsymbol{Y}} - \overline{\boldsymbol{X}}\hat{\boldsymbol{\beta}} + \overline{\boldsymbol{X}}(\hat{\boldsymbol{\beta}} - \hat{\boldsymbol{\beta}}_0)\right]^{\mathrm{T}}\left[\overline{\boldsymbol{Y}} - \overline{\boldsymbol{X}}\hat{\boldsymbol{\beta}} + \overline{\boldsymbol{X}}(\hat{\boldsymbol{\beta}} - \hat{\boldsymbol{\beta}}_0)\right] \\
&= \mathrm{RSS}(H_1) + I_1 + I_2 + I_3,
\end{aligned}$$

其中

$$\begin{aligned}
I_1 &= (\hat{\boldsymbol{\beta}} - \hat{\boldsymbol{\beta}}_0)^{\mathrm{T}}\overline{\boldsymbol{X}}^{\mathrm{T}}\overline{\boldsymbol{X}}(\hat{\boldsymbol{\beta}} - \hat{\boldsymbol{\beta}}_0), \\
I_2 &= (\hat{\boldsymbol{\beta}} - \hat{\boldsymbol{\beta}}_c)^{\mathrm{T}}\overline{\boldsymbol{X}}^{\mathrm{T}}\left(\overline{\boldsymbol{Y}} - \overline{\boldsymbol{X}}\hat{\boldsymbol{\beta}}\right), \\
I_3 &= \left(\overline{\boldsymbol{Y}} - \overline{\boldsymbol{X}}\hat{\boldsymbol{\beta}}\right)^{\mathrm{T}}\overline{\boldsymbol{X}}(\hat{\boldsymbol{\beta}} - \hat{\boldsymbol{\beta}}_0).
\end{aligned}$$

通过简单地计算, 可得

$$I_1 = \hat{\boldsymbol{\beta}}^{\mathrm{T}}\boldsymbol{A}^{\mathrm{T}}\left[\boldsymbol{A}(\overline{\boldsymbol{X}}^{\mathrm{T}}\overline{\boldsymbol{X}})^{-1}\boldsymbol{A}^{\mathrm{T}}\right]^{-1}\boldsymbol{A}\hat{\boldsymbol{\beta}}, \quad I_2 = I_3 = 0.$$

由定理 7.1, 可得

$$\mathrm{RSS}(H_0) - \mathrm{RSS}(H_1) \xrightarrow{\mathrm{D}} \sigma^2\chi_k^2(\lambda).$$

再根据 Slutsky 定理, 可得

$$2T_n = \frac{\mathrm{RSS}(H_0) - \mathrm{RSS}(H_1)}{\mathrm{RSS}(H_1)/n} \xrightarrow{\mathrm{D}} \chi_k^2(\lambda).$$

第 8 章　部分线性可加变量含误差模型的经验似然推断

近年来, 经验似然方法作为一类有效的非参数方法被应用到多类模型的研究中. 本章主要考虑利用经验似然方法来构造部分线性可加测量误差模型中参数分量的区间估计. 本章的主要内容来自文献 Wei,Luo 和 Wu (2012).

8.1　介　　绍

考虑如下的部分线性可加变量含误差模型

$$\begin{cases} Y = \boldsymbol{X}^{\mathrm{T}}\boldsymbol{\beta} + m_1(Z_1) + \cdots + m_q(Z_q) + \varepsilon, \\ \boldsymbol{V} = \boldsymbol{X} + \boldsymbol{\xi}, \end{cases} \tag{8.1}$$

其中 Y 为因变量, $\boldsymbol{X}$ 为 p 维自变量, $\boldsymbol{\beta} = (\beta_1, \beta_2, \cdots, \beta_p)^{\mathrm{T}}$ 为未知的参数分量, Z_k 为自变量, $m_1(\cdot), \cdots, m_q(\cdot)$ 是未知的非参数函数, 模型误差 ε 均值为零, 方差为 σ^2. 为了模型可识别, 假设 $E\{m_k(Z_k)\} = 0,\ k = 1, 2, \cdots, q$. 模型误差 $\boldsymbol{\xi}$ 独立于 $(Y, \boldsymbol{X}, Z_1, \cdots, Z_q)$, 均值为零, 协方差矩阵为 $\boldsymbol{\Sigma}_{\boldsymbol{\xi}}$. 为了讨论方便, 假设 $\boldsymbol{\Sigma}_{\boldsymbol{\xi}}$ 已知. 如果 $\boldsymbol{\Sigma}_{\boldsymbol{\xi}}$ 未知, 可以通过 $\boldsymbol{V}$ 的重复观测来构造其估计量.

显然, 模型 (8.1) 包含了很多常见的模型. 当协变量 $\boldsymbol{X}$ 可以精确观测时, 模型 (8.1) 即为 Opsomer 和 Ruppert (1999), Li (2000), Manzana 和 Zerom (2005), Jiang 等 (2007) 等文献中讨论的部分线性可加模型. 如果 $\boldsymbol{\beta} = 0$, 模型 (8.1) 转换为标准的可加模型. 如果 $q = 1$, 模型 (8.1) 则为部分线性测量误差模型.

对于模型 (8.1), 参数分量 $\boldsymbol{\beta}$ 往往是讨论的重点. Liang 等 (2008) 最早研究了该模型, 利用校正 profile 最小二乘方法提出了参数分量 $\boldsymbol{\beta}$ 的 $\sqrt{n}$ 相合估计量, 并证明了估计量的渐近正态性.

对于参数分量 $\boldsymbol{\beta}$, 构造其区间估计是必要和有意义的. 我们可以利用 Liang 等 (2008) 的结果采用正态逼近方法来构造参数分量的区间估计, 但是构造过程中涉及到复杂的方差估计. 为了避免这一问题, 本章将采用 Owen (1988,1990) 提出的经验似然方法来构造参数分量的区间估计. 作为一种有效的非参数方法, 经验似然具有很多优点并被广泛应用到各类模型和多种问题中, 详细的讨论可参考文献 Chen (1994), Kolaczyk (1994), Owen (1991), Qin 和 Lawless (1994) 和 Wang 和 Jing (1999)

等以及关于经验似然的著作 Owen (2001).

下面介绍 Liang 等 (2008) 所提出的校正 profile 最小二乘估计. 首先为了叙述的方便, 假设模型 (8.1) 中 $q=2$. 假设 $\{Y_i, \boldsymbol{X}_i, Z_{1i}, Z_{2i},\}_{i=1}^n$ 来自于模型 (8.1), 则有

$$Y_i = \boldsymbol{X}_i^{\mathrm{T}}\boldsymbol{\beta} + m_1(Z_{1i}) + m_2(Z_{2i}) + \varepsilon_i, \quad i=1,2,\cdots,n, \tag{8.2}$$

假设 $\boldsymbol{\beta}$ 已知, 则模型 (8.2) 可记为

$$Y_i - \boldsymbol{X}_i^{\mathrm{T}}\boldsymbol{\beta} = m_1(Z_{1i}) + m_2(Z_{2i}) + \varepsilon_i, \quad i=1,2,\cdots,n. \tag{8.3}$$

模型 (8.3) 是 Opsomer 和 Ruppert (1997) 研究过的两变量可加模型. 记

$$\boldsymbol{Y} = \begin{pmatrix} Y_1 \\ Y_2 \\ \vdots \\ Y_n \end{pmatrix}, \quad \boldsymbol{X} = \begin{pmatrix} \boldsymbol{X}_1^{\mathrm{T}} \\ \boldsymbol{X}_2^{\mathrm{T}} \\ \vdots \\ \boldsymbol{X}_n^{\mathrm{T}} \end{pmatrix}, \quad \boldsymbol{m}_1 = \begin{pmatrix} m_1(Z_{11}) \\ m_1(Z_{12}) \\ \vdots \\ m_1(Z_{1n}) \end{pmatrix},$$

$$\boldsymbol{m}_2 = \begin{pmatrix} m_2(Z_{21}) \\ m_2(Z_{22}) \\ \vdots \\ m_2(Z_{2n}) \end{pmatrix}, \quad \boldsymbol{\varepsilon} = \begin{pmatrix} \varepsilon_1 \\ \varepsilon_2 \\ \vdots \\ \varepsilon_n \end{pmatrix},$$

模型 (8.2) 可写为如下的矩阵形式

$$\boldsymbol{Y} = \boldsymbol{X}\boldsymbol{\beta} + \boldsymbol{m}_1 + \boldsymbol{m}_2 + \boldsymbol{\varepsilon}. \tag{8.4}$$

我们将使用 backfitting 方法估计可加模型 (8.3) 中的未知函数 $\{m_k(\cdot), k=1,2\}$. 定义针对第 k 个非参数函数的局部线性光滑矩阵为

$$\boldsymbol{S}_k = \begin{pmatrix} \boldsymbol{e}_1^{\mathrm{T}}\{\boldsymbol{D}_{Z_{k1}}^{k\mathrm{T}}\boldsymbol{K}_{Z_{k1}}\boldsymbol{D}_{Z_{k1}}^{k}\}^{-1}\boldsymbol{D}_{Z_{k1}}^{\mathrm{T}}\boldsymbol{K}_{Z_{k1}} \\ \boldsymbol{e}_1^{\mathrm{T}}\{\boldsymbol{D}_{Z_{k2}}^{k\mathrm{T}}\boldsymbol{K}_{Z_{k2}}\boldsymbol{D}_{Z_{k2}}^{k}\}^{-1}\boldsymbol{D}_{Z_{k2}}^{\mathrm{T}}\boldsymbol{K}_{Z_{k2}} \\ \vdots \\ \boldsymbol{e}_1^{\mathrm{T}}\{\boldsymbol{D}_{Z_{kn}}^{k\mathrm{T}}\boldsymbol{K}_{Z_{kn}}\boldsymbol{D}_{Z_{kn}}^{k}\}^{-1}\boldsymbol{D}_{Z_{kn}}^{\mathrm{T}}\boldsymbol{K}_{Z_{kn}} \end{pmatrix},$$

其中

$$\boldsymbol{D}_{Z_k}^{k} = \begin{pmatrix} 1 & (Z_{k1}-Z_k)/h_k \\ 1 & (Z_{k2}-Z_k)/h_k \\ \vdots & \vdots \\ 1 & (Z_{kn}-Z_k)/h_k \end{pmatrix}.$$

和 $\boldsymbol{e}_1=(1,0)^{\mathrm{T}}$, $\boldsymbol{K}_{Z_k}=\mathrm{diag}(K_{h_k}(Z_{k1}-Z_k),K_{h_k}(Z_{k2}-Z_k),\cdots,K_{h_k}(Z_{kn}-Z_k))$, 其中 $K_{h_k}(\cdot)=K(\cdot/h_k)/h_k$, $K(\cdot)$ 为核函数, h_k 为窗宽.

由 Hastie 和 Tibshirani (1990), 未知非参数函数 $\boldsymbol{m}_k$ 可以通过求解如下方程

$$\begin{pmatrix} \boldsymbol{I}_n & \boldsymbol{S}_1^* \\ \boldsymbol{S}_2^* & \boldsymbol{I}_n \end{pmatrix}\begin{pmatrix} \boldsymbol{m}_1 \\ \boldsymbol{m}_2 \end{pmatrix}=\begin{pmatrix} \boldsymbol{S}_1^* \\ \boldsymbol{S}_2^* \end{pmatrix}(\boldsymbol{Y}-\boldsymbol{X}\boldsymbol{\beta}),$$

其中 $\boldsymbol{S}_k^*=(\boldsymbol{I}_n-\boldsymbol{1}\boldsymbol{1}^{\mathrm{T}})\boldsymbol{S}_k, k=1,2$. 由 Opsomer 和 Ruppert (1997) 可知, $\boldsymbol{m}_1$ 和 $\boldsymbol{m}_2$ 的 backfitting 估计为

$$\hat{\boldsymbol{m}}_1=\boldsymbol{W}_1(\boldsymbol{Y}-\boldsymbol{X}\boldsymbol{\beta}),\quad \hat{\boldsymbol{m}}_2=\boldsymbol{W}_2(\boldsymbol{Y}-\boldsymbol{X}\boldsymbol{\beta}),\tag{8.5}$$

其中

$$\boldsymbol{W}_1=\boldsymbol{I}_n-(\boldsymbol{I}_n-\boldsymbol{S}_1^*\boldsymbol{S}_2^*)^{-1}(\boldsymbol{I}_n-\boldsymbol{S}_1^*),\quad \boldsymbol{W}_2=\boldsymbol{I}_n-(\boldsymbol{I}_n-\boldsymbol{S}_2^*\boldsymbol{S}_1^*)^{-1}(\boldsymbol{I}_n-\boldsymbol{S}_2^*).$$

将 $\hat{\boldsymbol{m}}_1$ 和 $\hat{\boldsymbol{m}}_2$ 代入模型 (8.2), 整理可得如下的线性模型

$$Y_i-\hat{Y}_i=(\boldsymbol{X}_i-\hat{\boldsymbol{X}}_i)^{\mathrm{T}}\boldsymbol{\beta}+\varepsilon_i-\hat{\varepsilon}_i,\quad i=1,2,\cdots,n,\tag{8.6}$$

其中 $\hat{\boldsymbol{Y}}=(\hat{Y}_1,\cdots,\hat{Y}_n)^{\mathrm{T}}=\boldsymbol{S}\boldsymbol{Y}$, $\hat{\boldsymbol{X}}=(\hat{\boldsymbol{X}}_1,\cdots,\hat{\boldsymbol{X}}_n)^{\mathrm{T}}=\boldsymbol{S}\boldsymbol{X}$, $\hat{\boldsymbol{\varepsilon}}=(\hat{\varepsilon}_1,\cdots,\hat{\varepsilon}_n)^{\mathrm{T}}=\boldsymbol{S}\boldsymbol{\varepsilon}$, 和 $\boldsymbol{S}=\boldsymbol{W}_1+\boldsymbol{W}_2$.

显然, 如果 $\boldsymbol{X}_i$ 可以精确观测, 那么基于模型 (8.6), 就可以得到 $\boldsymbol{\beta}$ 的 profile 最小二乘估计. 然而, $\boldsymbol{X}_i$ 在此是不能精确观测的. Liang 等 (2008) 构造 $\boldsymbol{\beta}$ 的校正 profile 最小二乘估计如下

$$\hat{\boldsymbol{\beta}}=\arg\min_{\boldsymbol{\beta}\in\mathbf{R}^p}\left[(\overline{\boldsymbol{Y}}-\overline{\boldsymbol{V}}\boldsymbol{\beta})^{\mathrm{T}}(\overline{\boldsymbol{Y}}-\overline{\boldsymbol{V}}\boldsymbol{\beta})-n\boldsymbol{\beta}^{\mathrm{T}}\boldsymbol{\Sigma}_{\boldsymbol{\xi}}\boldsymbol{\beta}\right]=(\overline{\boldsymbol{V}}^{\mathrm{T}}\overline{\boldsymbol{V}}-n\boldsymbol{\Sigma}_{\boldsymbol{\xi}})^{-1}\overline{\boldsymbol{V}}^{\mathrm{T}}\overline{\boldsymbol{Y}},\tag{8.7}$$

其中 $\overline{\boldsymbol{Y}}=\boldsymbol{Y}-\hat{\boldsymbol{Y}},\overline{\boldsymbol{V}}=\boldsymbol{V}-\hat{\boldsymbol{V}}$, $\hat{\boldsymbol{V}}=(\hat{\boldsymbol{V}}_1,\cdots,\hat{\boldsymbol{V}}_n)^{\mathrm{T}}=\boldsymbol{S}\boldsymbol{V}$ 和 $\boldsymbol{V}=(\boldsymbol{V}_1,\cdots,\boldsymbol{V}_n)^{\mathrm{T}}$.

8.2 参数分量的经验似然

Liang 等 (2008) 针对参数分量 $\boldsymbol{\beta}$ 所构造的校正最小二乘估计是下面估计方程的解,

$$\frac{1}{n}\sum_{i=1}^{n}\boldsymbol{W}_{in}(\boldsymbol{\beta})=\boldsymbol{0},\tag{8.8}$$

其中

$$\boldsymbol{W}_{in}(\boldsymbol{\beta})=(\boldsymbol{V}_i-\hat{\boldsymbol{V}}_i)\left\{Y_i-\hat{Y}_i-(\boldsymbol{V}_i-\hat{\boldsymbol{V}}_i)^{\mathrm{T}}\boldsymbol{\beta}\right\}+\boldsymbol{\Sigma}_{\boldsymbol{\xi}}\boldsymbol{\beta},\quad i=1,2,\cdots,n,$$

以及 $\hat{\boldsymbol{V}}=(\hat{\boldsymbol{V}}_1,\cdots,\hat{\boldsymbol{V}}_n)^{\mathrm{T}}=\boldsymbol{S}\boldsymbol{V}$, $\boldsymbol{V}=(\boldsymbol{V}_1,\cdots,\boldsymbol{V}_n)^{\mathrm{T}}$. 由 Owen(2001), $\boldsymbol{\beta}$ 的经验似然函数可定义为

$$l_n(\boldsymbol{\beta})=-2\max\left\{\sum_{i=1}^n\log(np_i):\sum_{i=1}^n p_i\boldsymbol{W}_{in}(\boldsymbol{\beta})=0,p_i\geqslant 0,\sum_{i=1}^n p_i=1\right\}. \tag{8.9}$$

由 Lagrange 乘子法, p_i 的最优解为

$$p_i=\frac{1}{n}\left(1+\boldsymbol{\lambda}^{\mathrm{T}}\boldsymbol{W}_{in}(\boldsymbol{\beta})\right)^{-1},\quad i=1,2,\cdots,n, \tag{8.10}$$

其中 $\boldsymbol{\lambda}=(\lambda_1,\lambda_2,\cdots,\lambda_p)^{\mathrm{T}}$ 是如下方程的解

$$\frac{1}{n}\sum_{i=1}^n\frac{\boldsymbol{W}_{in}(\boldsymbol{\beta})}{1+\boldsymbol{\lambda}^{\mathrm{T}}\boldsymbol{W}_{in}(\boldsymbol{\beta})}=0, \tag{8.11}$$

则对数经验似然函数为

$$l_n(\boldsymbol{\beta})=2\sum_{i=1}^n\log\left(1+\boldsymbol{\lambda}^{\mathrm{T}}\boldsymbol{W}_{in}(\boldsymbol{\beta})\right). \tag{8.12}$$

下面给出一些假设条件. 这些条件在 Liang 等 (2008) 中也使用过.

条件 1　$E(\varepsilon|\boldsymbol{X},Z_1,Z_2)=0$ 和 $E(|\varepsilon|^3|\boldsymbol{X},Z_1,Z_2)<\infty$.

条件 2　窗宽 $h_1,h_2\sim n^{-1/5}$.

条件 3　核函数 $K(\cdot)$ 为对称密度函数, 具有紧支撑, 满足 $\int K(u)\mathrm{d}u=1$, $\int K(u)u\mathrm{d}u=0$ 和 $\int u^2K(u)\mathrm{d}u=1$. 非参数函数 $m_k(\cdot),k=1,2$ 在其支撑上连续可微.

条件 4　Z_1 和 Z_2 的密度函数都大于 0, 且有有界的二阶连续导数.

条件 5　$E(\|\boldsymbol{\xi}\|^3)<\infty$.

下面的定理说明我们所构造的经验对数似然比函数 $l_n(\boldsymbol{\beta})$ 渐近分布为标准的 χ^2 分布.

定理 8.1　若上面的条件 1～ 条件 5 成立, 如果 $\boldsymbol{\beta}_0$ 是参数 $\boldsymbol{\beta}$ 的真实值, 有

$$l_n(\boldsymbol{\beta}_0)\xrightarrow{D}\chi_p^2,$$

其中 χ_p^2 为自由度为 p 的 χ^2 分布.

利用定理 2.1 可以构造参数分量 $\boldsymbol{\beta}$ 的区间估计. 定义 $\boldsymbol{I}_\alpha(\boldsymbol{\beta})=\{\boldsymbol{\beta}:l_n(\boldsymbol{\beta})\leqslant c_\alpha\}$, 其中 c_α 是 χ_p^2 $(1-\alpha)$ 分位数. $\boldsymbol{I}_\alpha(\boldsymbol{\beta})$ 就是 $\boldsymbol{\beta}$ 渐近水平为 $1-\alpha$ 置信区域, 即有

$$P(\boldsymbol{\beta}_0\in\boldsymbol{I}_\alpha(\boldsymbol{\beta}))=1-\alpha+o(1).$$

注 8.1 如果 $\boldsymbol{\Sigma}_{\boldsymbol{\xi}}$ 未知, 常用的方法就是利用重复观测, 假设有 $\boldsymbol{V}_{ij}=\boldsymbol{X}_i+\boldsymbol{\xi}_{ij}, j=1,2,\cdots,l_i$. 为了方便, 简化为 $l_i\equiv 2$. 定义

$$\overline{\boldsymbol{V}}_i=(\boldsymbol{V}_{i1}+\boldsymbol{V}_{i2})/2,\quad \hat{\boldsymbol{\Sigma}}_{\boldsymbol{\xi}}=n^{-1}\sum_{i=1}^{n}\sum_{j=1}^{2}(\boldsymbol{V}_i-\overline{\boldsymbol{V}}_i)(\boldsymbol{V}_i-\overline{\boldsymbol{V}}_i)^{\mathrm{T}}.$$

很容易证明 $\hat{\boldsymbol{\Sigma}}_{\boldsymbol{\xi}}$ 是 $\boldsymbol{\Sigma}_{\boldsymbol{\xi}}$ 的无偏估计. 因此, 利用 $\overline{\boldsymbol{V}}_i$ 代替 $\boldsymbol{V}_i$, 用 $\hat{\boldsymbol{\Sigma}}_{\boldsymbol{\xi}}\boldsymbol{\beta}$ 代替 $\boldsymbol{\Sigma}_{\boldsymbol{\xi}}\boldsymbol{\beta}$, 定理 8.1 仍然成立.

注 8.2 对于 $q>2$ 时的模型 (8.1), 针对参数分量所构造经验对数似然比函数的方法和上述 $q=2$ 时的一样, 只不过将光滑矩阵 $\boldsymbol{S}$ 替换成针对含有 q 个非参数函数的可加结构 $\boldsymbol{m}=\boldsymbol{m}_1+\cdots+\boldsymbol{m}_q$ 的光滑函数. 详细的讨论可参考 Opsomer (2000).

注 8.3 实际数据分析中, 我们经常遇到因变量存在缺失的情况. Wei, Jia 和 Hu (2013b) 从理论上探讨了部分线性可加变量含误差模型 (8.1) 在因变量缺失下的估计以及经验似然推断问题.

8.3 数值模拟

本节将通过数值模拟来比较所提出的经验似然方法和正态逼近方法.

由 Liang 等 (2008) 中的定理 1, 容易证明

$$\sqrt{n}\hat{\boldsymbol{\Omega}}^{-1/2}\hat{\boldsymbol{\Sigma}}(\hat{\boldsymbol{\beta}}-\boldsymbol{\beta})\xrightarrow{D}N(\boldsymbol{0},\boldsymbol{I}_p),$$

其中 $\hat{\boldsymbol{\Sigma}}=\dfrac{1}{n}\sum_{i=1}^{n}(\boldsymbol{V}_i-\hat{\boldsymbol{V}}_i)(\boldsymbol{V}_i-\hat{\boldsymbol{V}}_i)^{\mathrm{T}}-\boldsymbol{\Sigma}_{\boldsymbol{\xi}}$ 和

$$\hat{\boldsymbol{\Omega}}=n^{-1}\sum_{i=1}^{n}\left\{(\boldsymbol{V}_i-\hat{\boldsymbol{V}}_i)[\boldsymbol{Y}_i-\hat{\boldsymbol{Y}}_i-(\boldsymbol{V}_i-\hat{\boldsymbol{V}}_i)^{\mathrm{T}}\hat{\boldsymbol{\beta}}]+\boldsymbol{\Sigma}_{\boldsymbol{\xi}}\hat{\boldsymbol{\beta}}\right\}^2.$$

因此, 基于正态逼近方法可构造 $\boldsymbol{\beta}$ 的置信水平为 $1-\alpha$ 的置信区间

$$\hat{\boldsymbol{\beta}}\pm z_{1-\alpha/2}/\left(\sqrt{n}\hat{\boldsymbol{\Omega}}^{-1/2}\hat{\boldsymbol{\Sigma}}\right),$$

其中 $z_{1-\alpha/2}$ 满足 $\Phi(z_{1-\alpha/2})=1-\alpha/2$.

假设数据产生于如下的部分线性可加模型

$$y_i=x_i\beta+m_1(z_{1i})+m_2(z_{2i})+\varepsilon_i,\quad v_i=x_i+e_i,\quad i=1,2,\cdots,n,$$

其中 $\beta=1, m_1(z_{1i})=\sin(2\pi z_{1i}), m_2(z_{2i})=z_{2i}^3+3z_{2i}^2-2z_{2i}-1, x_i\sim N(0,1), e_i\sim N(0,0.1^2), z_{1i}\sim U(0,1)$, 和 $z_{2i}\sim U(0,1)$. 为了考察误差分布对最终结果的影

响, 分别取 $\varepsilon_i \sim N(0,1)$ 和 $\varepsilon_i \sim U(-\sqrt{3},\sqrt{3})$. 核函数 $K(x)=0.75(1-x^2)\boldsymbol{I}_{|x|\leqslant 1}$, 为了方便, 取窗宽 $h_1=h_2=n^{-1/5}$.

对上述每种情况, 基于样本 n 重复试验 1000 次, 分别基于正态逼近方法 (NA) 和所提的经验似然方法 (EL) 构造对应于 $\alpha=0.05$ 的置信区间, 计算区间长度和覆盖率, 结果见表 8.1.

表 8.1 $\boldsymbol{\beta}$ 置信水平为 0.95 的置信区间的覆盖 (CP) 和平均长度 (AL)

	n	CP(EL)	CP(NA)	AL(EL)	AL(NA)
$\varepsilon_i \sim N(0,1)$	30	0.903	0.891	0.7169	0.7201
	60	0.932	0.922	0.5230	0.5169
	80	0.945	0.937	0.4506	0.4437
	100	0.941	0.931	0.4015	0.3950
$\varepsilon_i \sim U(-\sqrt{3},\sqrt{3})$	30	0.914	0.906	0.7119	0.7239
	60	0.930	0.909	0.5147	0.5118
	80	0.944	0.928	0.4445	0.4402
	100	0.950	0.929	0.3972	0.3971

从表 8.1 中不难看出, 各种情况下经验似然方法得到的置信区间覆盖了要高于正态逼近方法所得到的置信区间覆盖率. 随着样本量的增大, 两种方法得到的置信区间覆盖率都越来越接近于 0.95, 置信区间的平均长度越来越小.

8.4 定理的证明

下面先给出几个引理.

引理 8.1 令 $(\boldsymbol{X}_1,\boldsymbol{Y}_1),\cdots,(\boldsymbol{X}_n,\boldsymbol{Y}_n)$ 为独立同分布 (iid) 的随机序列, 其中 $\boldsymbol{Y}_i, i=1,2,\cdots,n$ 为一元随机变量, 进一步假定 $E|y|^s<\infty$ 与 $\sup_x \int |y|^s f(x,y)\mathrm{d}y<\infty$, 其中 f 表示 $(\boldsymbol{X},\boldsymbol{Y})$ 的联合密度. 令 K 为一有界正函数, 并有有界支撑且满足 Lipschitz 条件, 则有

$$\sup_x\left|\frac{1}{n}\sum_{i=1}^{n}\left[K_h(\boldsymbol{X}_i-x)\boldsymbol{Y}_i-E\left(K_h(\boldsymbol{X}_i-x)\boldsymbol{Y}_i\right)\right]\right|=O_p\left(\left\{\frac{\log(1/h)}{nh}\right\}^{1/2}\right),$$

其中对于 $\varepsilon<1-s^{-1}$ 有 $n^{2\varepsilon-1}h\to\infty$.

该引理可由 Mack 和 Silverman (1982) 的结论得到.

引理 8.2 若条件 1~ 条件 5 成立, 则有

$$\max_{1\leqslant i\leqslant n}\|\boldsymbol{W}_{in}(\boldsymbol{\beta})\|=o_p(n^{1/2}).$$

证明 由 Opsomer 和 Ruppert (1997) 的引理 3.1 和引理 3.2, 有

$$\boldsymbol{S}_1^* = \boldsymbol{S}_1 - \mathbf{1}_n\mathbf{1}_n^{\mathrm{T}}/n + o_p(\mathbf{1}_n\mathbf{1}_n^{\mathrm{T}}/n), \quad (\boldsymbol{I}_n - \boldsymbol{S}_1^*\boldsymbol{S}_2^*)^{-1} = \boldsymbol{I}_n + o_p(1/n).$$

由引理 8.1, 可证明

$$\boldsymbol{SV} = \begin{bmatrix} E(\boldsymbol{X}_1|Z_{11}) + E(\boldsymbol{X}_1|Z_{21}) \\ E(\boldsymbol{X}_2|Z_{12}) + E(\boldsymbol{X}_2|Z_{22}) \\ \vdots \\ E(\boldsymbol{X}_n|Z_{1n}) + E(\boldsymbol{X}_n|Z_{2n}) \end{bmatrix} \{1 + O_p(c_n)\},$$

$$\boldsymbol{S}(\boldsymbol{m}_1 + \boldsymbol{m}_2) = \begin{bmatrix} m_1(Z_{11}) + m_2(Z_{21}) \\ m_1(Z_{12}) + m_2(Z_{22}) \\ \vdots \\ m_1(Z_{1n}) + m_2(Z_{2n}) \end{bmatrix} \{1 + O_p(c_n)\},$$

和 $\boldsymbol{S}(\boldsymbol{\varepsilon} - \boldsymbol{U\beta}) = \mathbf{1}_n O_p(d_n)$, $c_n = h_1^2 + h_2^2 + \left\{\dfrac{\log(1/h_1)}{nh_1}\right\}^{1/2} + \left\{\dfrac{\log(1/h_2)}{nh_2}\right\}^{1/2}$, $d_n = \left(\dfrac{\log n}{nh_1}\right)^{1/2} + \left(\dfrac{\log n}{nh_2}\right)^{1/2}$, $\boldsymbol{\varepsilon} = (\varepsilon_1, \cdots, \varepsilon_n)^{\mathrm{T}}$, 以及 $\boldsymbol{U} = (\boldsymbol{\xi}_1, \cdots, \boldsymbol{\xi}_n)^{\mathrm{T}}$. 由以上结论, 有

$$\boldsymbol{V}_i - \hat{\boldsymbol{V}}_i = \boldsymbol{V}_i - [E(\boldsymbol{X}_i|Z_{1i}) + E(\boldsymbol{X}_i|Z_{2i})] + o_p(1), \tag{8.13}$$

和

$$Y_i - \hat{Y}_i - (\boldsymbol{V}_i - \hat{\boldsymbol{V}}_i)^{\mathrm{T}}\boldsymbol{\beta} = \varepsilon_i - \boldsymbol{\xi}_i^{\mathrm{T}}\boldsymbol{\beta} + o_p(1). \tag{8.14}$$

由 Owen (1990) 的引理 3, 可得

$$\max_{1\leqslant i\leqslant n} \|\boldsymbol{V}_i\| = o_p(n^{1/2}), \quad \max_{1\leqslant i\leqslant n} |\varepsilon_i| = o_p(n^{1/2}), \quad \max_{1\leqslant i\leqslant n} |\boldsymbol{\xi}_i^{\mathrm{T}}\boldsymbol{\beta}| = o_p(n^{1/2}). \tag{8.15}$$

由式 (8.13)~ 式 (8.15), 可得

$$\max_{1\leqslant i\leqslant n} \|\boldsymbol{W}_{in}(\boldsymbol{\beta})\| \leqslant \max_{1\leqslant i\leqslant n} \|\boldsymbol{V}_i - \hat{\boldsymbol{V}}_i\| \times \max_{1\leqslant i\leqslant n} |Y_i - \hat{Y}_i - (\boldsymbol{V}_i - \hat{\boldsymbol{V}}_i)^{\mathrm{T}}\boldsymbol{\beta}| + o_p(n^{1/2}) = o_p(n^{1/2}).$$

引理 8.3 若条件 1~ 条件 5 成立, 有

$$\frac{1}{\sqrt{n}}\sum_{i=1}^n \boldsymbol{W}_{in} \xrightarrow{D} N(\mathbf{0}, \boldsymbol{\Omega}), \tag{8.16}$$

$$\frac{1}{n}\sum_{i=1}^n \boldsymbol{W}_{in}\boldsymbol{W}_{in}^{\mathrm{T}} \xrightarrow{p} \boldsymbol{\Omega}, \tag{8.17}$$

其中 $\boldsymbol{\Omega} = E\left\{\overline{\boldsymbol{X}}_1(\varepsilon_1 - \boldsymbol{\xi}_1^{\mathrm{T}}\boldsymbol{\beta})\right\}^{\otimes 2} + E(\varepsilon_1^2\boldsymbol{\xi}_1\boldsymbol{\xi}_1^{\mathrm{T}}) + E\left\{[(\boldsymbol{\xi}_1\boldsymbol{\xi}_1^{\mathrm{T}} - \boldsymbol{\Sigma}_{\boldsymbol{\xi}})\boldsymbol{\beta}]^{\otimes 2}\right\}$, $\overline{\boldsymbol{X}}_i = \boldsymbol{X}_i - E(\boldsymbol{X}_i|Z_{1i}) - E(\boldsymbol{X}_i|Z_{2i})$, $i = 1, 2$, $\boldsymbol{A}^{\otimes 2} = \boldsymbol{A}\boldsymbol{A}^{\mathrm{T}}$.

证明 由 $\boldsymbol{W}_{in}$ 的定义, 可得

$$\begin{aligned}\frac{1}{\sqrt{n}}\sum_{i=1}^{n}\boldsymbol{W}_{in} &= \frac{1}{\sqrt{n}}\boldsymbol{V}^{\mathrm{T}}(\boldsymbol{I}-\boldsymbol{S})^{\mathrm{T}}(\boldsymbol{I}-\boldsymbol{S})(\boldsymbol{Y}-\boldsymbol{V}\boldsymbol{\beta}) + \sqrt{n}\boldsymbol{\Sigma}_{\boldsymbol{\xi}}\boldsymbol{\beta} \\ &= \frac{1}{\sqrt{n}}\boldsymbol{V}^{\mathrm{T}}(\boldsymbol{I}-\boldsymbol{S})^{\mathrm{T}}(\boldsymbol{I}-\boldsymbol{S})(\boldsymbol{m}_1+\boldsymbol{m}_2) \\ &\quad + \frac{1}{\sqrt{n}}\boldsymbol{V}^{\mathrm{T}}(\boldsymbol{I}-\boldsymbol{S})^{\mathrm{T}}(\boldsymbol{I}-\boldsymbol{S})(\boldsymbol{\varepsilon}-\boldsymbol{U}\boldsymbol{\beta}) \\ &\quad + \frac{1}{\sqrt{n}}\sum_{i=1}^{n}\boldsymbol{\Sigma}_{\boldsymbol{\xi}}\boldsymbol{\beta}.\end{aligned}$$

根据 Liang 等 (2008) 的定理 1 有

$$\frac{1}{n}\boldsymbol{V}^{\mathrm{T}}(\boldsymbol{I}-\boldsymbol{S})^{\mathrm{T}}(\boldsymbol{I}-\boldsymbol{S})(\boldsymbol{m}_1+\boldsymbol{m}_2) = O_p(n^{-1/2}h_1^2 + n^{-1/2}h_2^2) + o_p(h_1^4 + h_2^4),$$

$$\frac{1}{n}\boldsymbol{V}^{\mathrm{T}}(\boldsymbol{I}-\boldsymbol{S})^{\mathrm{T}}(\boldsymbol{I}-\boldsymbol{S})(\boldsymbol{\varepsilon}-\boldsymbol{U}\boldsymbol{\beta}) = \frac{1}{n}\sum_{i=1}^{n}(\overline{\boldsymbol{X}}_i + \boldsymbol{\xi}_i)(\varepsilon_i - \boldsymbol{\xi}_i^{\mathrm{T}}\boldsymbol{\beta}) + O_p(n^{-1/2}h_1^2 + n^{-1/2}h_2^2).$$

结合条件 1 和条件 2, 有

$$\frac{1}{\sqrt{n}}\sum_{i=1}^{n}\boldsymbol{W}_{in} = \frac{1}{\sqrt{n}}\sum_{i=1}^{n}\left[(\overline{\boldsymbol{X}}_i + \boldsymbol{\xi}_i)(\varepsilon_i - \boldsymbol{\xi}_i^{\mathrm{T}}\boldsymbol{\beta}) + \boldsymbol{\Sigma}_{\boldsymbol{\xi}}\boldsymbol{\beta}\right] + o_p(1). \tag{8.18}$$

记 $\boldsymbol{\zeta}_i = (\overline{\boldsymbol{X}}_i + \boldsymbol{\xi}_i)(\varepsilon_i - \boldsymbol{\xi}_i^{\mathrm{T}}\boldsymbol{\beta}) + \boldsymbol{\Sigma}_{\boldsymbol{\xi}}\boldsymbol{\beta}$. 显然, $\boldsymbol{\zeta}_i$ 为独立同分布的随机向量, 并有 $E(\boldsymbol{\zeta}_i) = \mathbf{0}$ 和 $\mathrm{Cov}(\boldsymbol{\zeta}_i) = \boldsymbol{\Omega}$. 由中心极限定理

$$\frac{1}{\sqrt{n}}\sum_{i=1}^{n}\boldsymbol{\zeta}_i \xrightarrow{D} N(\mathbf{0}, \boldsymbol{\Omega}). \tag{8.19}$$

该结论再加上式 (8.18) 可证式 (8.16) 成立.

类似于引理 8.2 和式 (8.16) 的证明, 可证式 (8.17) 成立. 在此省略.

引理 8.4 若条件 1~ 条件 5 成立, 有 $\|\boldsymbol{\lambda}\| = O_p(n^{-1/2})$.

证明 记 $\boldsymbol{\lambda} = \rho\boldsymbol{\theta}$, 其中 $\rho \geqslant 0$, $\boldsymbol{\theta} \in \mathbf{R}^p$ 和 $\|\boldsymbol{\theta}\| = 1$. 类似于 Owen (1991), 有

$$\begin{aligned}0 = \frac{1}{n}\boldsymbol{\theta}^{\mathrm{T}}\sum_{i=1}^{n}\frac{\boldsymbol{W}_{in}}{1+\rho\boldsymbol{\theta}^{\mathrm{T}}\boldsymbol{W}_{in}} \leqslant& \frac{1}{n}\boldsymbol{\theta}^{\mathrm{T}}\sum_{i=1}^{n}\boldsymbol{W}_{in} - \frac{\rho}{n(1+\rho\max\limits_{1\leqslant i\leqslant n}\|\boldsymbol{W}_{in}\|)} \\ &\times \mathrm{mineig}\left(\frac{1}{n}\sum_{i=1}^{n}\boldsymbol{W}_{in}\boldsymbol{W}_{in}^{\mathrm{T}}\right),\end{aligned}$$

其中 mineig($\boldsymbol{Q}$) 是矩阵 $\boldsymbol{Q}$ 的最小特征根. 由此可得

$$\rho\,\text{mineig}\left(\frac{1}{n}\sum_{i=1}^{n}\boldsymbol{W}_{in}\boldsymbol{W}_{in}^{\mathrm{T}}\right)-\frac{\rho}{n}\boldsymbol{\theta}^{\mathrm{T}}\sum_{i=1}^{n}\boldsymbol{W}_{in}\max_{1\leqslant i\leqslant n}\|\boldsymbol{W}_{in}\|\leqslant\frac{1}{n}\boldsymbol{\theta}^{\mathrm{T}}\sum_{i=1}^{n}\boldsymbol{W}_{in}.$$

基于引理 8.3, 有

$$\frac{1}{n}\boldsymbol{\theta}^{\mathrm{T}}\sum_{i=1}^{n}\boldsymbol{W}_{in}=O_p(n^{-1/2}),\quad \text{mineig}\left(\frac{1}{n}\sum_{i=1}^{n}\boldsymbol{W}_{in}\boldsymbol{W}_{in}^{\mathrm{T}}\right)=O_p(1).$$

由上可得 $\rho=O_p(n^{-1/2})$, 从而 $\|\boldsymbol{\lambda}\|=O_p(n^{-1/2})$.

定理 8.1 的证明 由引理 8.2 和引理 8.4, 可得

$$\max_{1\leqslant i\leqslant n}|\boldsymbol{\lambda}^{\mathrm{T}}\boldsymbol{W}_{in}|=O_p(n^{-1/2})o_p(n^{1/2})=o_p(1).$$

进一步由 Taylor 展开可得

$$l_n(\boldsymbol{\beta}_0)=2\sum_{i=1}^{n}\log(1+\boldsymbol{\lambda}^{\mathrm{T}}\boldsymbol{W}_{in})=2\sum_{i=1}^{n}\left(\boldsymbol{\lambda}^{\mathrm{T}}\boldsymbol{W}_{in}-\frac{1}{2}(\boldsymbol{\lambda}^{\mathrm{T}}\boldsymbol{W}_{in})^2\right)+r_n,\tag{8.20}$$

其中

$$P\left(|r_n|\leqslant C\sum_{i=1}^{n}(\boldsymbol{\lambda}^{\mathrm{T}}\boldsymbol{W}_{in})^3\right)\to 1,\quad n\to\infty.$$

由引理 8.2~ 引理 8.4, 可得

$$|r_n|\leqslant C\|\boldsymbol{\lambda}\|^3\max_{i}\|\boldsymbol{W}_{in}\|\sum_{i=1}^{n}\|\boldsymbol{W}_{in}\|^2=o_p(1).\tag{8.21}$$

注意到

$$0=\sum_{i=1}^{n}\frac{\boldsymbol{W}_{in}}{1+\boldsymbol{\lambda}^{\mathrm{T}}\boldsymbol{\xi}_{in}}=\sum_{i=1}^{n}\boldsymbol{W}_{in}-\left(\sum_{i=1}^{n}\boldsymbol{W}_{in}\boldsymbol{W}_{in}^{\mathrm{T}}\right)\boldsymbol{\lambda}+\sum_{i=1}^{n}\frac{\boldsymbol{W}_{in}(\boldsymbol{\lambda}^{\mathrm{T}}\boldsymbol{W}_{in})^2}{1+\boldsymbol{\lambda}^{\mathrm{T}}\boldsymbol{W}_{in}}.\tag{8.22}$$

再基于引理 8.2~ 引理 8.4, 可得

$$\boldsymbol{\lambda}=\left(\sum_{i=1}^{n}\boldsymbol{W}_{in}\boldsymbol{W}_{in}^{\mathrm{T}}\right)^{-1}\sum_{i=1}^{n}\boldsymbol{W}_{in}+o_p(n^{-1/2}).\tag{8.23}$$

式 (8.11) 两边同乘以 $\boldsymbol{\lambda}^{\mathrm{T}}$, 有

$$0=\sum_{i=1}^{n}\frac{\boldsymbol{\lambda}^{\mathrm{T}}\boldsymbol{W}_{in}}{1+\boldsymbol{\lambda}^{\mathrm{T}}\boldsymbol{W}_{in}}=\sum_{i=1}^{n}(\boldsymbol{\lambda}^{\mathrm{T}}\boldsymbol{W}_{in})-\sum_{i=1}^{n}(\boldsymbol{\lambda}^{\mathrm{T}}\boldsymbol{W}_{in})^2+\sum_{i=1}^{n}\frac{(\boldsymbol{\lambda}^{\mathrm{T}}\boldsymbol{W}_{in})^3}{1+\boldsymbol{\lambda}^{\mathrm{T}}\boldsymbol{W}_{in}}.\tag{8.24}$$

由引理 8.2~ 引理 8.4, 可得

$$\sum_{i=1}^{n}\frac{(\boldsymbol{\lambda}^{\mathrm{T}}\boldsymbol{W}_{in})^3}{1+\boldsymbol{\lambda}^{\mathrm{T}}\boldsymbol{W}_{in}}=o_p(1). \tag{8.25}$$

基于式 (8.24) 和式 (8.25), 有

$$\sum_{i=1}^{n}\boldsymbol{\lambda}^{\mathrm{T}}\boldsymbol{W}_{in}=\sum_{i=1}^{n}\left(\boldsymbol{\lambda}^{\mathrm{T}}\boldsymbol{W}_{in}\right)^2+o_p(1). \tag{8.26}$$

由式 (8.20), (8.21), (8.23) 和 (8.25), 可得

$$\begin{aligned} l_n(\boldsymbol{\beta}_0)&=\sum_{i=1}^{n}\boldsymbol{\lambda}^{\mathrm{T}}\boldsymbol{W}_{in}\boldsymbol{W}_{in}^{\mathrm{T}}\boldsymbol{\lambda}+o_p(1)\\ &=\left(\frac{1}{\sqrt{n}}\sum_{i=1}^{n}\boldsymbol{W}_{in}\right)^{\mathrm{T}}\left(\frac{1}{n}\sum_{i=1}^{n}\boldsymbol{W}_{in}\boldsymbol{W}_{in}^{\mathrm{T}}\right)^{-1}\left(\frac{1}{\sqrt{n}}\sum_{i=1}^{n}\boldsymbol{W}_{in}\right)+o_p(1). \end{aligned}$$

最后, 由引理 8.3, 有 $l_n(\boldsymbol{\beta}_0)\xrightarrow{D}\chi_p^2, n\to\infty$.

第 9 章　利用部分线性模型检验线性回归关系

近年来, 借助于非参数回归技术检验参数回归关系的方法以其备择假设的广泛性已越来越受到重视. 然而, “维数祸根”问题使得多种检验方法在自变量为多个的情形时的使用受到限制. 为了克服这一局限性, 本章研究利用部分线性模型检验线性关系. 首先构造了广义似然比检验统计量, 并证明了该统计量在原假设下的渐近分布为 χ^2 分布; 并讨论了误差为正态分布时检验统计量的构造以及检验 p 值的计算; 其次基于导函数的性质提出了一种新的检验方法; 最后一部分对于部分线性模型给出了一种新的估计 profile 局部加权最小二乘估计, 并基于该估计讨论了相关的检验问题. 本章内容主要来自于文献魏传华和吴喜之 (2007, 2008d) 以及魏传华, 李静和吴喜之 (2009).

9.1　引　　言

实际数据分析中, 很多问题涉及变量之间关系的研究, 参数回归模型作为解决此类问题的有效途径而得到了广泛的研究, 它以其计算简单、理论完整而常常作为人们拟合数据的首选模型之一. 假设自变量 X 与因变量 Y 满足回归关系

$$m(x) = E(Y|X = x) \quad \text{或者记为} \quad Y = m(x) + \varepsilon.$$

参数回归分析中, 假定均值函数 $m(\cdot)$ 属于某一参数族 $\mu = \{m(\cdot;\boldsymbol{\theta});\boldsymbol{\theta} \in \boldsymbol{\Theta}\}$, 其中 $\boldsymbol{\Theta}$ 为 p 维 Euclidean 空间的一个子集, p 有限. 根据上面的设定, 可以基于观测数据 $(Y_i, X_i)_{i=1}^n$ 进行未知参数 $\boldsymbol{\theta}$ 的估计以及其他方面的统计推断. 但是, 当实际数据的生成机制与设定的参数回归模型有较大偏差时, 这样的分析往往会得出与实际不符, 甚至错误的结论. 因此, 在利用参数回归模型解决实际问题时, 必须要检验设定的回归函数能否很好地解释因变量与自变量之间的关系, 即要研究如下的假设检验问题

$$H_0 : m \in \mu \quad \text{VS} \quad H_1 : m \in \mu. \tag{9.1}$$

对于上面的检验问题, 传统的做法往往是基于残差图进行判断, 即判断残差或者残差的某种函数形式 (如学生化残差) 对于因变量拟合值或者某个自变量有无趋势性. 显然这种检验方法有很强的主观性, 很多细微的趋势只通过图形可能是观察不出来的. 另外一种常用的方法就是在线性模型中添加自变量的高阶项及交叉乘积

项并借助于各模型的拟合优度考察非线性关系的存在性, 但该方法忽视了与这些高次项正交方向上的非线性性, 而且这样选择的非线性关系有很大的局限性. 随着计算机计算能力的飞速发展, 非参数回归方法近三十年来得到广泛的研究. 近些年来, 借助于非参数回归技术检验参数回归关系的方法以其备择假设的广泛性已越来越受到重视, 目前这方面已取得了相当丰富的研究成果. 关于这些检验方法的总结与比较可参考 Hart(1997), Pagan 和 Ullah(1999), Miles 和 Mora(2003), Yatchew(2003), Zhang 和 Dette(2004). 下面列出几种常用的检验方法.

第一类方法的基本思想是以非参数回归模型为备择假设, 利用非参数光滑方法估计回归函数, 然后基于某种距离比较回归函数的非参数拟合值 $\hat{m}(\cdot)$ 与参数方法拟合值 $\hat{m}(\cdot,\hat{\boldsymbol{\theta}})$, 显然二者的差距越大, 越倾向于拒绝原假设, 即认为参数模型 $\hat{m}(\cdot,\boldsymbol{\theta})$ 不能很好的反映因变量与自变量之间的关系. 根据距离的设定, 该类检验统计量主要有两种形式. 一种是两种拟合值的直接比较, 检验统计量主要是下面的形式

$$T_{1n}=\int\{\hat{m}_h(x)-\hat{g}_h(x;\boldsymbol{\theta})\}^2\pi_1(x)\mathrm{d}x,$$

Hardle 和 Mammen (1993), Gonzalez-Manteiga 和 Cao (1993), Alcala, Cristobal 和 Gonzalez-Manteiga (1999) 都属于该方法. 另外一种类似于经典回归分析中的拟合优度检验 (如 F 检验), 是对两种拟合方法的残差平方和进行比较, 即为如下形式

$$T_{2n}=\sum_{i=1}^{n}\{Y_i-g(X_i;\boldsymbol{\theta})\}^2-\sum_{i=1}^{n}\{Y_i-\hat{m}_h(X_i)\}^2.$$

Ullah (1985) 最早提出利用 T_{2n}/n 基于核光滑方法来检验参数回归模型, Fan 和 Li (2002) 从理论上研究了检验统计量 T_{2n}/n 的渐近性质, 另外, Dette (1999) 提出的方差估计之差检验统计量以及 Fan,Zhang 和 Zhang (2001) 提出的广义似然比检验统计量都属于该种方法.

第二类检验方法是基于残差分析的思想. 这其中包括 Hart (1997) 所说的光滑残差法 (smoothing residuals), Yatchew (2003) 所定义的残差回归方法 (residual regression) 以及 Azzalini 和 Bowman(1993) 与 Dette (2000) 研究的伪似然比方法 (pseudolikelihood ratio test).

第三类方法是条件矩检验统计量, 主要是 Zheng (1996), Li 和 Wang(1998), Liu, Stengos 和 Li (2000), G. Ellison 和 S. Ellison (2001) 研究的如下形式的统计量

$$T_{3n}=\sum\sum_{1\leqslant i\neq j\leqslant n}K_h(X_i-X_j)\{Y_i-g(X_i;\boldsymbol{\theta})\}\{Y_j-g(X_j;\boldsymbol{\theta})\}.$$

上面的这几类检验方法主要是基于核 (局部多项式) 光滑技术, 基于其他光滑方法的特点来构造检验统计量也得到了人们的重视, 主要包括 Yanagimoto 和 Yanagimoto (1987), Cox 等 (1988), Eubank 和 Spiegelman(1990) 以及 Jayasuriya (1996)

等利用光滑样条方法检验参数回归关系. Hong 和 White (1995) 基于级数估计构造了检验统计量. 由于这些方法的检验统计量一般都较为复杂, 其零分布很难精确求得, 通常利用统计量的极限分布或 bootstrap 再抽样方法等模拟统计量的零分布以确定检验的临界值或 p 值.

然而, 由于非参数回归方法在处理高维数据时会遇到"维数祸根"问题, 所以上面的许多方法的研究都是基于自变量 X 为一维的情形. 但是, 实际数据分析中遇到最多的往往是有多个自变量的情形. 为了克服这一局限性, 下面研究基于部分线性模型来检验线性回归关系.

部分线性模型结构如下

$$Y = f(T) + \boldsymbol{X}^{\mathrm{T}}\boldsymbol{\beta} + \varepsilon, \tag{9.2}$$

其中 $(T, \boldsymbol{X}^{\mathrm{T}})$ 是自变量, 不失一般性, 下面的叙述中假定 T 为一维变量, Y 为因变量,ε 为模型误差, 有 $E(\varepsilon) = 0$ 和 $\mathrm{Var}(\varepsilon) = \sigma^2$, $\boldsymbol{\beta} = (\beta_1, \beta_2, \cdots, \beta_p)^{\mathrm{T}}$ 为 p 维未知待估参数, $f(\cdot)$ 为一未知函数. 该模型自从 Engle 等 (1986) 提出以后受到了人们的广泛关注, 在理论与应用上都得到了较为深入的研究, 详细讨论可参考 Heckman (1986), Chen (1988), Robinson (1988), Speckman (1988), Andrews (1992), Yatchew (1997), Hardle, Liang 和 Gao (2000). 然而, 文献中的大部分工作都侧重于模型的估计, 关于模型的检验问题目前还没有得到很好的研究. 我们感兴趣的是以部分线性模型做为备择假设来检验线性回归关系. 不失一般性, 考虑如下的假设检验问题

$$H_0: \quad f(T) = \theta_0 + \theta_1 T + \cdots + \theta_k T^k \quad \text{VS} \quad H_1: \quad f(\cdot)\text{为其他类型函数}. \tag{9.3}$$

对于该类检验问题, Fan 和 Huang (2001) 提出了 Adaptive Neyman 检验, Gonzalez-Manteiga 和 Aneiros-Rerez (2003) 提出了 Cramer-von-Mises 类型的检验统计量, Crainiceanu 等 (2005) 基于惩罚样条 (penalised splines) 方法构造了似然比与限制似然比检验统计量.

在众多非参数检验方法中, Fan, Zhang 和 Zhang (2001) 提出的广义似然比检验方法最近受到了人们的重视. 一方面该检验方法适用范围广, Fan, Zhang 和 Zhang (2001) 研究了高斯白噪声模型 (gaussian white noise model)、非参数回归模型、变系数模型, Fan 和 Huang (2005) 研究了部分线性变系数模型, Fan 和 Jiang (2005) 研究了可加模型. 另一方面是上面的文献都证明了该统计量在原假设成立的渐近分布为 χ^2 分布, 即具有 Wilks 现象. 虽然部分线性变系数模型是部分线性模型的推广, 但由于部分线性模型结构的特殊性, Fan 和 Huang (2005) 的某些结果并不能直接用到上面关于 $f(\cdot)$ 的检验问题上. 下面首先针对问题 (9.3) 构造广义似然比检验统计量, 并给出其原假设成立时的渐近分布.

9.2　广义似然比检验方法

对于部分线性模型 (9.2), 已经有多种方法提出以估计未知参数 $\boldsymbol{\beta}$ 和未知函数 $f(\cdot)$, 其中一个应用非常广泛的方法为 profile 最小二乘估计. 另外鉴于局部线性方法在非参数回归模型估计中的优良性, 本书下面对于模型 (9.2) 采用基于局部线性方法的 profile 最小二乘估计方法.

首先假定对于模型 (9.2) 有 n 次随机观测样本为 $\{(T_i, X_{i1}, \cdots, X_{ip}, y_i)\}$, $i = 1, 2, \cdots, n$. 假定模型 (9.2) 中 $(\beta_1, \beta_2, \cdots, \beta_p)^{\mathrm{T}}$ 已知 (给定), 则模型 (9.2) 转化为如下形式的非参数回归模型

$$Y_i^* = f(T_i) + \varepsilon_i, \quad i = 1, 2, \cdots, n, \tag{9.4}$$

其中 $Y_i^* = Y_i - (X_{i1}\beta_1 + \cdots + X_{ip}\beta_p)$. 对于非参数回归模型 (9.4), 利用局部多项式方法来估计 $f(\cdot)$. 对 T_0 附近的一点 T, 对 $f(T)$ 利用 Taylor 展开式有

$$f(T) \approx f(T_0) + f'(T_0)(T - T_0) + \cdots + \frac{f^{(k)}(T_0)}{k!}(T - T_0)^k,$$

从而可通过针对 $\alpha_0(T_0), \cdots, \alpha_k(T_0)$ 极小化

$$\sum_{i=1}^{n} \left[Y_i^* - \alpha_0(T_0) - \alpha_1(T_0)(T_i - T_0) - \cdots - \alpha_k(T_0)(T_i - T_0)^k\right]^2 K_h(T_i - T_0), \tag{9.5}$$

得到它们的估计 $\hat{\alpha}_0(T_0), \cdots, \hat{\alpha}_k(T_0)$. 其中对于 $j = 0, 1, \cdots, k$, 有 $\alpha_j(\cdot) = f^{(j)}(\cdot)/j!$, $K_h(\cdot) = K(\cdot/h)/h$, K 是核函数, h 是窗宽.

为了叙述方便, 采用下面的矩阵形式, 记

$$\boldsymbol{Y} = \begin{pmatrix} y_1 \\ y_2 \\ \vdots \\ y_n \end{pmatrix}, \quad \boldsymbol{X} = \begin{pmatrix} \boldsymbol{X}_1^{\mathrm{T}} \\ \boldsymbol{X}_2^{\mathrm{T}} \\ \vdots \\ \boldsymbol{X}_n^{\mathrm{T}} \end{pmatrix} = \begin{pmatrix} X_{11} & \cdots & X_{1p} \\ X_{21} & \cdots & X_{2p} \\ \vdots & & \vdots \\ X_{n1} & \cdots & X_{np} \end{pmatrix}, \quad \boldsymbol{f} = \begin{pmatrix} f(T_1) \\ f(T_2) \\ \vdots \\ f(T_n) \end{pmatrix}$$

与

$$\boldsymbol{\varepsilon} = \begin{pmatrix} \varepsilon_1 \\ \varepsilon_2 \\ \vdots \\ \varepsilon_n \end{pmatrix}, \quad \boldsymbol{D}_{\mathrm{T}_0} = \begin{pmatrix} 1 & T_1 - T_0 & \cdots & (T_1 - T_0)^k \\ 1 & T_2 - T_0 & \cdots & (T_2 - T_0)^k \\ \vdots & \vdots & & \vdots \\ 1 & T_n - T_0 & \cdots & (T_n - T_0)^k \end{pmatrix},$$

以及

$$\boldsymbol{W}_{u_0} = \mathrm{diag}(K_h(T_1 - T_0), K_h(T_2 - Y_0), \cdots, K_h(T_n - T_0)).$$

从而模型 (9.4) 可写为如下的矩阵形式

$$\boldsymbol{Y}-\boldsymbol{X}\boldsymbol{\beta}=\boldsymbol{f}+\boldsymbol{\varepsilon}. \tag{9.6}$$

基于问题 (9.5), 由广义最小二乘法可得

$$(\hat{\alpha}(T_0),\cdots,\hat{\alpha}_k(T_0))^{\mathrm{T}}=\{\boldsymbol{D}_{T_0}^{\mathrm{T}}\boldsymbol{W}_{T_0}\boldsymbol{D}_{T_0}\}^{-1}\boldsymbol{D}_{T_0}^{\mathrm{T}}\boldsymbol{W}_{T_0}(\boldsymbol{Y}-\boldsymbol{X}\boldsymbol{\beta}), \tag{9.7}$$

则 $\boldsymbol{f}$ 的初次估计为

$$\tilde{\boldsymbol{f}}=\boldsymbol{S}(\boldsymbol{Y}-\boldsymbol{X}\boldsymbol{\beta}),$$

其中

$$\boldsymbol{S}=\begin{pmatrix}\boldsymbol{e}_{1,k+1}^{\mathrm{T}}\{\boldsymbol{D}_{T_1}^{\mathrm{T}}\boldsymbol{W}_{T_1}\boldsymbol{D}_{T_1}\}^{-1}\boldsymbol{D}_{T_1}^{\mathrm{T}}\boldsymbol{W}_{T_1}\\ \boldsymbol{e}_{1,k+1}^{\mathrm{T}}\{\boldsymbol{D}_{T_2}^{\mathrm{T}}\boldsymbol{W}_{T_2}\boldsymbol{D}_{T_2}\}^{-1}\boldsymbol{D}_{T_2}^{\mathrm{T}}\boldsymbol{W}_{T_2}\\ \vdots\\ \boldsymbol{e}_{1,k+1}^{\mathrm{T}}\{\boldsymbol{D}_{T_n}^{\mathrm{T}}\boldsymbol{W}_{T_n}\boldsymbol{D}_{T_n}\}^{-1}\boldsymbol{D}_{T_n}^{\mathrm{T}}\boldsymbol{W}_{T_n}\end{pmatrix},$$

其中 $\boldsymbol{e}_{j,k+1}, j=1,\cdots,k+1$ 表示长度为 $k+1$ 的单位向量, 其中第 j 个元素为 1, 其余为零. 若无特殊说明, 下同. 将 $\boldsymbol{f}$ 的初次估计代入式 (9.6), 整理可得如下的线性回归模型

$$\overline{\boldsymbol{Y}}=\overline{\boldsymbol{X}}\boldsymbol{\beta}+\boldsymbol{\varepsilon}, \tag{9.8}$$

其中

$$\overline{\boldsymbol{Y}}=(\overline{Y}_1,\overline{Y}_2,\cdots,\overline{Y}_n)^{\mathrm{T}}=(\boldsymbol{I}-\boldsymbol{S})\boldsymbol{Y},\quad \overline{\boldsymbol{X}}=(\overline{\boldsymbol{X}}_1,\overline{\boldsymbol{X}}_2,\cdots,\overline{\boldsymbol{X}}_n)^{\mathrm{T}}=(\boldsymbol{I}-\boldsymbol{S})\boldsymbol{X}.$$

利用最小二乘法估计上面的模型, 则得 $\boldsymbol{\beta}$ 的 profile 最小二乘估计为

$$\hat{\boldsymbol{\beta}}=[\boldsymbol{X}^{\mathrm{T}}(\boldsymbol{I}-\boldsymbol{S})^{\mathrm{T}}(\boldsymbol{I}-\boldsymbol{S})\boldsymbol{X}]^{-1}\boldsymbol{X}^{\mathrm{T}}(\boldsymbol{I}-\boldsymbol{S})^{\mathrm{T}}(\boldsymbol{I}-\boldsymbol{S})\boldsymbol{Y}. \tag{9.9}$$

最后得 $\boldsymbol{f}$ 的最终估计为

$$\hat{\boldsymbol{f}}=\boldsymbol{S}(\boldsymbol{Y}-\boldsymbol{X}\hat{\boldsymbol{\beta}}).$$

从而 $\boldsymbol{Y}$ 的拟合值为

$$\hat{\boldsymbol{Y}}=\hat{\boldsymbol{f}}+\boldsymbol{X}\hat{\boldsymbol{\beta}}=\boldsymbol{L}\boldsymbol{Y}, \tag{9.10}$$

其中

$$\boldsymbol{L}=\boldsymbol{S}+(\boldsymbol{I}-\boldsymbol{S})\boldsymbol{X}[\boldsymbol{X}^{\mathrm{T}}(\boldsymbol{I}-\boldsymbol{S})^{\mathrm{T}}(\boldsymbol{I}-\boldsymbol{S})\boldsymbol{X}]^{-1}\boldsymbol{X}^{\mathrm{T}}(\boldsymbol{I}-\boldsymbol{S})^{\mathrm{T}}(\boldsymbol{I}-\boldsymbol{S}).$$

对于非参数部分的线性检验问题 (9.3), 类似于 Fan 和 Huang (2005), 假定 $\boldsymbol{\varepsilon}\sim N(\boldsymbol{0},\sigma^2\boldsymbol{I}_n)$, 显然模型 (9.2) 的对数似然函数为

$$L(\boldsymbol{f},\boldsymbol{\beta},\sigma^2)=-\frac{n}{2}\log 2\pi\sigma^2-\sum_{i=1}^{n}\frac{(Y_i-f(T_i)-\boldsymbol{X}_i^{\mathrm{T}}\boldsymbol{\beta})^2}{2\sigma^2}.$$

给定 $\boldsymbol{\beta}$, 那么 $f(\cdot)$ 可以利用局部线性方法得到其估计为 $\hat{f}(\cdot;\boldsymbol{\beta})$, 将该估计代入上面的似然函数, 则得到如下的 profile 似然函数

$$L(\hat{\boldsymbol{\alpha}}(\cdot;\boldsymbol{\beta}),\boldsymbol{\beta},\sigma^2)=-\frac{n}{2}\log 2\pi\sigma^2-\sum_{i=1}^{n}\frac{(\tilde{Y}_i-\tilde{\boldsymbol{X}}_i^{\mathrm{T}}\boldsymbol{\beta})^2}{2\sigma^2},$$

则由 $\dfrac{\partial L(\boldsymbol{\alpha},\boldsymbol{\beta},\sigma^2)}{\partial\boldsymbol{\beta}}=0$ 可得 $\boldsymbol{\beta}$ 的 Proflile 似然估计为式 (9.9) 的 $\hat{\boldsymbol{\beta}}$, 由 $\dfrac{\partial L(\boldsymbol{\alpha},\boldsymbol{\beta},\sigma^2)}{\partial\sigma^2}=0$ 得 σ^2 的估计为 $\hat{\sigma}^2=\mathrm{RSS}_1/n$, 其中

$$\mathrm{RSS}_1=\sum_{i=1}^{n}\left\{Y_i-\hat{f}(T_i)-\boldsymbol{X}_i^{\mathrm{T}}\hat{\boldsymbol{\beta}}\right\}^2.$$

将 $\boldsymbol{\beta},\sigma^2$ 的估计代入上面 profile 似然函数, 整理可得

$$L(H_1)=-\frac{n}{2}-\frac{n}{2}\log(2\pi/n)-\frac{n}{2}\log\mathrm{RSS}_1. \tag{9.11}$$

当原假设成立时, 模型 (9.2) 为如下的线性回归模型,

$$Y=\theta_0+\theta_1T+\cdots+\theta_kT^k+\boldsymbol{X}\boldsymbol{\beta}.$$

由线性模型的极大似然估计理论可知, 最后得对数似然函数为

$$L(H_0)=-\frac{n}{2}-\frac{n}{2}\log(2\pi/n)-\frac{n}{2}\log\mathrm{RSS}_0, \tag{9.12}$$

其中

$$\mathrm{RSS}_0=\sum_{i=1}^{n}\left\{Y_i-\overline{\theta}_0-\cdots-\overline{\theta}_kT_i^k-\boldsymbol{X}_i^{\mathrm{T}}\overline{\boldsymbol{\beta}}\right\}^2$$

中间的 $(\overline{\theta}_0,\cdots,\overline{\theta}_k,\overline{\beta}_1,\cdots,\overline{\beta}_p)$ 为 $(\theta_0,\cdots,\theta_k,\beta_1,\cdots,\beta_p)$ 的极大似然估计.

基于上面的似然函数, 构造广义似然比检验检验量

$$T_{\mathrm{GLR}}=L(H_1)-L(H_0)=\frac{n}{2}\log\frac{\mathrm{RSS}_0}{\mathrm{RSS}_1}\approx\frac{n}{2}\frac{\mathrm{RSS}_0-\mathrm{RSS}_1}{\mathrm{RSS}_1}. \tag{9.13}$$

对于检验统计量 T_{GLR} 有如下结论.

定理 9.1 如果原假设 H_0^N 成立, 且满足 9.5 节中的条件, 则有当 $k=1$ 时

$$r_kT_{\mathrm{GLR}}\sim\chi^2_{\delta_n},$$

其中

$$r_k=\frac{K(0)-\dfrac{1}{2}\displaystyle\int K^2(t)\mathrm{d}t}{\displaystyle\int(K(t)-\frac{1}{2}K*K(t))^2\mathrm{d}t},$$

$|\boldsymbol{\Omega}|$ 为变量 T 支撑的长度,

$$\delta_n = r_k \frac{|\boldsymbol{\Omega}|}{h}\left(K(0) - \frac{1}{2}\int K^2(t)\mathrm{d}t\right).$$

若 $k \geqslant 1$, 将上面 r_k 和 δ_n 中的核函数 $K(\cdot)$ 都换成等价核函数 (详见 Fan 和 Gijbels(1996)[64]), 上面结论仍然成立.

定理 9.1 说明了对于部分线性模型非参数部分的检验, 广义似然比检验统计量具有很好的性质, 即存在 Wilks 现象. 从而该定理也是对广义似然比检验的一个推广.

注 9.1 显然上面的检验包含了关于变量 T 的显著性检验问题 $H_0: f(T) = \alpha$, 以及变量 T 的线性假设检验问题 $H_0: f(T) = \alpha_0 + \alpha_1 T$.

注 9.2 上面的检验中, 假定了模型是同方差的, 现在将结果推广到一类异方差情形. 对于模型误差 ε, 假定 $E(\varepsilon|T=t, \boldsymbol{X}=\boldsymbol{x}) = 0, E(\varepsilon^2|T=t, \boldsymbol{X}=\boldsymbol{x}) = \sigma^2(t)$, 此时利用加权残差平方和

$$\mathrm{RSS}_0 = \sum_{i=1}^{n}\left\{Y_i - \overline{\theta}_0 - \cdots - \overline{\theta}_k T_i^k - \boldsymbol{X}_i^{\mathrm{T}}\overline{\boldsymbol{\beta}}\right\}^2 w(T_i),$$

$$\mathrm{RSS}_1 = \sum_{i=1}^{n}\left\{Y_i - \hat{f}(T_i) - \boldsymbol{X}_i^{\mathrm{T}}\hat{\boldsymbol{\beta}}\right\}^2 w(T_i).$$

如果权函数 $w(\cdot)$ 在 $\{t: g(t) > 0\}$ 上的紧支撑上光滑, 则对于检验问题 H_0^N, 定理 1 的结果可推广如下

$$r_k' T_N \sim \chi^2_{\delta_n'}, \tag{9.14}$$

其中

$$r_k' = r_k[E\sigma^2(T)w(T)]\int \sigma^2(t)w(t)\mathrm{d}t\left[\sigma^4(t)w^2(t)\mathrm{d}t\right]^{-1},$$

$$\delta_n' = r_k h^{-1}\left(K(0) - \frac{1}{2}\int K^2(t)\mathrm{d}t\right)\left[\int \sigma^2(t)w(t)\mathrm{d}t\right]^2\left[\int \sigma^4(t)w^2(t)\mathrm{d}t\right]^{-1}.$$

如果 $\sigma^2(t) = v(t)\sigma^2$, $v(t)$ 为一已知函数, 则令 $w(t) = v(t)^{-1}$, 则有 $r_k' = r_k, \delta_n' = \delta_n$, 即定理依然成立.

上面的广义似然比统计量虽然是在误差为正态分布下构造的, 但是结果的证明并没有依靠这个假设. 下面研究误差为正态分布时针对检验问题 H_0 的检验统计量的构造以及检验 p 值的计算.

类似于线性回归模型的 F 检验, 构造如下的检验统计量

$$T_1 = \frac{\mathrm{RSS}_0 - \mathrm{RSS}_1}{\mathrm{RSS}_1}. \tag{9.15}$$

显然由式 (9.12) 知

$$\mathrm{RSS}_1 = \|\boldsymbol{Y} - \hat{\boldsymbol{Y}}\|^2 = \boldsymbol{Y}^{\mathrm{T}}(\boldsymbol{I} - \boldsymbol{L})^{\mathrm{T}}(\boldsymbol{I} - \boldsymbol{L})\boldsymbol{Y}.$$

若原假设成立, 由线性回归模型的理论可知

$$\mathrm{RSS}_0 = \boldsymbol{\varepsilon}^{\mathrm{T}}\left[\boldsymbol{I} - \boldsymbol{Z}(\boldsymbol{Z}^{\mathrm{T}}\boldsymbol{Z})^{-1}\boldsymbol{Z}^{\mathrm{T}}\right]\boldsymbol{\varepsilon},$$

其中

$$\boldsymbol{Z} = \begin{pmatrix} 1 & T_1 & \cdots & T_1^k & x_{11} & \cdots & x_{1p} \\ 1 & T_2 & \cdots & T_2^k & x_{21} & \cdots & x_{2p} \\ \vdots & \vdots & & \vdots & \vdots & & \vdots \\ 1 & T_n & \cdots & T_n^k & x_{n1} & \cdots & x_{np} \end{pmatrix}.$$

另外, 对于模型 (9.2) 的 profile 最小二乘估计, 有如下性质.

定理 9.2 如果原假设 H_0^N 成立, 部分线性模型 (9.2) 的 profile 最小二乘估计满足

$$\text{(a)}\quad E\hat{\boldsymbol{\beta}} = \boldsymbol{\beta}, \quad E\hat{f}(\mathrm{T}_i) = f(T_i) = \theta_0 + \theta_1 T_i + \cdots + \theta_k T_i^k,$$

对于残差平方和 RSS_1 有

$$\text{(b)}\quad \mathrm{RSS}_1 = \boldsymbol{\varepsilon}^{\mathrm{T}}(\boldsymbol{I} - \boldsymbol{L})^{\mathrm{T}}(\boldsymbol{I} - \boldsymbol{L})\boldsymbol{\varepsilon}.$$

由上面的结论有

$$F = \frac{\mathrm{RSS}_0 - \mathrm{RSS}_1}{\mathrm{RSS}_1} = \frac{\boldsymbol{\varepsilon}^{\mathrm{T}}\left\{\boldsymbol{I} - \boldsymbol{Z}(\boldsymbol{Z}^{\mathrm{T}}\boldsymbol{Z})^{-1}\boldsymbol{Z}^{\mathrm{T}} - (\boldsymbol{I} - \boldsymbol{L})^{\mathrm{T}}(\boldsymbol{I} - \boldsymbol{L})\right\}\boldsymbol{\varepsilon}}{\boldsymbol{\varepsilon}^{\mathrm{T}}(\boldsymbol{I} - \boldsymbol{L})^{\mathrm{T}}(\boldsymbol{I} - \boldsymbol{L})\boldsymbol{\varepsilon}}. \tag{9.16}$$

显然这类形式的统计量正是在第 1 章讨论的常见的正态分布二次型之比形式, 如果假设 F_N 的观测值为 f, 则由上式整理可得

$$p_0 = P_{H_0}(F > f) = P\{\boldsymbol{\varepsilon}^{\mathrm{T}}\boldsymbol{W}\boldsymbol{\varepsilon} > 0\}, \tag{9.17}$$

其中 $\boldsymbol{W} = \boldsymbol{Q}_0 - f\boldsymbol{Q}_1$, $\boldsymbol{Q}_0 = \boldsymbol{I} - \boldsymbol{Z}(\boldsymbol{Z}^{\mathrm{T}}\boldsymbol{Z})^{-1}\boldsymbol{Z}^{\mathrm{T}} - (\boldsymbol{I} - \boldsymbol{L})^{\mathrm{T}}(\boldsymbol{I} - \boldsymbol{L})$, $\boldsymbol{Q}_1 = (\boldsymbol{I} - \boldsymbol{L})^{\mathrm{T}}(\boldsymbol{I} - \boldsymbol{L})$. 基于结论 1.1~ 结论 1.3 得到下面的结果.

定理 9.3 设模型 (9.2) 中误差项 $\varepsilon_1, \varepsilon_2, \cdots, \varepsilon_n$ 为独立同分布的随机变量, 均服从均值为零, 方差为 σ^2 的正态分布,

(1) 若利用 1.3 节中的精确方法, 则 p_0 可由式 (1.11)~ 式 (1.13) 计算, 其中 $\lambda_1, \lambda_2, \cdots, \lambda_m$ 为矩阵 $\boldsymbol{W}$ 的互不相同的非零特征值, 而 $h_1, h_2, \cdots, h_m$ 分别为 $\lambda_1, \lambda_2, \cdots, \lambda_m$ 的重数.

(2) 若利用 1.3 节中的三阶矩 χ^2 逼近方法, 则当 $\mathrm{tr}(\boldsymbol{W}^3) > 0$ 时,

$$p_0 = P_{H_0}(F > f) \approx P(\chi_d^2 > d - h),$$

当 $\mathrm{tr}(\boldsymbol{W}^3) < 0$ 时,

$$p_0 = P_{H_0}(F > f) \approx P(\chi_d^2 < d - h),$$

其中 χ_d^2 是自由度为 d 的 χ^2 变量, 且

$$\begin{cases} d = \dfrac{\{\mathrm{tr}(\boldsymbol{W}^2)\}^3}{\{\mathrm{tr}(\boldsymbol{W}^3)\}^2}, \\ h_i = \dfrac{\mathrm{tr}(\boldsymbol{W}^2)\ \mathrm{tr}(\boldsymbol{W})}{\mathrm{tr}(\boldsymbol{W}^3)}. \end{cases}$$

(3) 若利用 1.3 节中的 F 分布逼近方法, 则

$$p_0 = P_{H_0}(F > f) \approx P\left(F(r_1, r_2) > \frac{\mathrm{tr}\boldsymbol{Q}_1}{\mathrm{tr}\boldsymbol{Q}_0} f\right),$$

其中 $r_1 = \dfrac{[\mathrm{tr}(\boldsymbol{Q}_0)]^2}{\mathrm{tr}(\boldsymbol{Q}_0^2)}$, $r_2 = \dfrac{[\mathrm{tr}(\boldsymbol{Q}_1)]^2}{\mathrm{tr}(\boldsymbol{Q}_q^2)}$, $F(r_1, r_2)$ 为服从自由度为 r_1 和 r_2 的 F 分布的随机变量.

定理 9.3 关于检验 p 值的计算依赖于误差为正态分布这个假定. 虽然本节以及很多文献通过数值模拟都验证了三阶矩 χ^2 逼近方法对于某些非正态分布误差具有一定的稳健性, 但是该方法对于误差的分布还是具有很强的局限性. 另外, 从理论上讲我们可以根据广义似然比检验统计量的渐近零分布求取检验的临界值, 但是 T_{GLR} 这类检验统计量的收敛速度很慢, 有限样本下利用渐近分布进行检验效果不是很好. 为了克服这些问题, 正如很多作者建议, 我们可以采用 bootstrap 方法基于检验统计量 F_k 或者 T_{GLR} 进行检验.

数值模拟 下面将通过模拟试验考察前面所提出的检验方法的有效性. 假设数据产生于如下的部分线性模型

$$y_i = f(t_i) + 2x_i + \varepsilon_i, \quad i = 1, 2, \cdots, n,$$

其中自变量 t_i 为固定设计点, 有 $t_i = \dfrac{i - 0.5}{n}$, 非参数函数为 $f(t_i) = -0.5 + c_1 t_i + c_2 t_i^2$, 自变量 $x_i = 5t_i^2 + \eta_i$, 其中 η_i 是服从 $U(-0.5, 0.5)$ 分布的随机数. 为了考察误差项服从正态分布的假定对检验功效的影响, 我们取误差项为正态分布 $N(0, 0.05^2)$ 和均匀分布 $U(-0.05\sqrt{3}, 0.05\sqrt{3})$ 两种情况, 这两种分布都满足均值为 0, 标准差为 0.05.

对上面两种误差分布, 分别取样本量 $n{=}100$ 以及 $(c_1, c_2) = (0, 0), (0, 0.5), (1, 0), (1, 0.5)$, 利用基于局部线性光滑技术的 profile 最小二乘方法估计上面的部分线性

模型. 依照本节的方法分别构造检验统计量 F_0, F_1, F_2 用以检验非参数函数 $f(t_i) = 1 + c_1 t_i + c_2 t_i^2$ 是否为常数、线性函数以及二次函数 (即针对 H_0 分别取 $k = 0, 1, 2$). 依照定理 9.3 利用三阶矩 χ^2 逼近方法求取检验 p- 值, 其中选取的核函数是 Epanechnikov 核 $K(u) = 0.75(1 - u^2)_+$. 为了解窗宽对检验功效的影响, 分别取 $h = 0.1, 0.3, 0.5, 0.7, 0.9, 1.1$ 进行模拟试验. 取显著水平 $\alpha = 0.05$, 对于上面的每种情况, 重复试验 500 次, 以 500 次重复中检验 p 值小于 α (即拒绝 H_0) 的频率模拟检验功效. 结果见表 9.1 和表 9.2.

表 9.1 $\varepsilon \sim N(0, 0.05^2)$ 时 500 次重复下拒绝原假设的频率

窗宽	$c_1 = c_2 = 0$			$c_1 = 0, c_2 = 0.5$			$c_1 = 1, c_2 = 0$			$c_1 = 1, c_2 = 0.5$		
	F_0	F_1	F_2	F_0	F_1	F_2	F_0	F_1	F_2	F_0	F_1	F_2
0.1	0.060	0.058	0.050	0.978	0.880	0.044	1.000	0.028	0.048	1.000	0.872	0.050
0.3	0.064	0.062	0.060	0.998	0.966	0.010	1.000	0.056	0.036	1.000	0.980	0.004
0.5	0.040	0.026	0.044	0.996	0.968	0.002	1.000	0.058	0.054	1.000	0.972	0.000
0.7	0.054	0.058	0.088	0.996	0.978	0.000	1.000	0.046	0.062	1.000	0.986	0.000
0.9	0.044	0.052	0.100	0.980	0.984	0.000	1.000	0.042	0.096	1.000	0.984	0.000
1.1	0.048	0.030	0.062	0.956	0.986	0.000	1.000	0.036	0.068	1.000	0.980	0.000

表 9.2 $\varepsilon \sim U(-0.05\sqrt{3}, 0.05\sqrt{3})$ 时 500 次重复下拒绝原假设的频率

窗宽	$c_1 = c_2 = 0$			$c_1 = 0, c_2 = 0.5$			$c_1 = 1, c_2 = 0$			$c_1 = 1, c_2 = 0.5$		
	F_1	F_2	F_3	F_1	F_2	F_3	F_1	F_2	F_3	F_1	F_2	F_3
0.1	0.042	0.050	0.056	0.984	0.882	0.050	1.000	0.050	0.038	1.000	0.904	0.052
0.3	0.062	0.040	0.032	0.996	0.968	0.010	1.000	0.066	0.054	1.000	0.982	0.022
0.5	0.050	0.044	0.044	0.996	0.978	0.000	1.000	0.040	0.046	1.000	0.978	0.000
0.7	0.038	0.056	0.078	0.998	0.976	0.000	1.000	0.064	0.064	1.000	0.990	0.000
0.9	0.042	0.044	0.090	0.988	0.988	0.000	1.000	0.038	0.110	1.000	0.988	0.000
1.1	0.046	0.048	0.058	0.964	0.988	0.000	1.000	0.062	0.040	1.000	0.994	0.000

由模拟结果可知:

(1) 虽然检验 p 值的计算公式是在误差项服从正态分布的假定下得到的. 但是, 对于均匀分布, 检验功效和精确性与正态分布的情况非常接近. 这说明本书所提检验方法的功效以及精确性关于误差项分布的变化具有一定的稳健性.

(2) 当原假设 H_0 为真时, 拒绝 H_0 的频率比较接近于显著水平 $\alpha = 0.05$, 即检验是比较可靠的. 当原假设不成立时, 拒绝 H_0 的频率一般接近 1.

(3) 当窗宽 h 在一定范围内变化时, 检验的功效是相当稳定的. 但从模拟中也发现, h 的一些取值使得检验的功效有所降低. 因此恰当的选取窗宽值是是非常重要的. 这也是利用非参数光滑技术检验参数回归关系中尚待研究的问题. 实用中, 可以用交叉证实法等选择方法确定的窗宽作为参考, 在一个更大的范围内考察检验

p 值随 h 的变化, 以更好地判断是否拒绝原假设.

9.3 基于导函数的非参数检验方法

最近, Gijbels 和 Rousson(2001) 提出了一种新的检验方法, 利用非参数回归技术检验某一函数是否为一特定阶数的多项式, 即为模型 (9.2) 中 $\boldsymbol{X}=0$ 时的检验问题 (9.3). 该方法最后将模型的检验转化为普通的线性回归模型中的检验问题, 从而可以直接利用 F 检验方法. 然而该方法中有几个关键的参数要选择, 而且最后的选择往往对结果产生很大的影响. 为了克服这一缺点, Mei 等 (2003) 提出了另外一种方法, 并给出了检验 p 值的计算方法. 下面基于该思想来研究部分线性模型的线性检验问题, 即模型 (9.2) 的检验问题 (9.3).

对于部分线性模型 (9.2), 依然利用前面介绍的基于局部多项式光滑的 profile 最小二乘估计方法来估计其中的未知函数 $f(\cdot)$ 和未知参数 $\boldsymbol{\beta}$. 当原假设 H_0 成立时, 该估计除了定理 9.2 的结果外, 还有下面的结论.

定理 9.4 如果原假设 H_0 成立, 则 $f^{(k)}(T_i)$ 的估计 $\hat{f}^{(k)}(t_i)$ 为无偏估计, 从而有

$$E\hat{\alpha}_k(T_i)=f^{(k)}(T_i)/k!=\theta_k.$$

由 9.2 节知道 $\alpha_k(T_i)$ 的估计为

$$\begin{aligned}\hat{\alpha}_k(T_i)=&\boldsymbol{e}_{k+1,k+1}^{\mathrm{T}}\{\boldsymbol{D}_{T_i}^{\mathrm{T}}\boldsymbol{W}_{T_i}\boldsymbol{D}_{T_i}\}^{-1}\boldsymbol{D}_{T_i}^{\mathrm{T}}\boldsymbol{W}_{T_i}(\boldsymbol{Y}-\boldsymbol{X}\hat{\boldsymbol{\beta}})\\=&\boldsymbol{e}_{k+1,k+1}^{\mathrm{T}}\{\boldsymbol{D}_{T_i}^{\mathrm{T}}\boldsymbol{W}_{T_i}\boldsymbol{D}_{T_i}\}^{-1}\boldsymbol{D}_{T_i}^{\mathrm{T}}\boldsymbol{W}_{T_i}\\&\left(\boldsymbol{I}-\boldsymbol{X}[\boldsymbol{X}^{\mathrm{T}}(\boldsymbol{I}-\boldsymbol{S})^{\mathrm{T}}(\boldsymbol{I}-\boldsymbol{S})\boldsymbol{X}]^{-1}\boldsymbol{X}^{\mathrm{T}}(\boldsymbol{I}-\boldsymbol{S})^{\mathrm{T}}(\boldsymbol{I}-\boldsymbol{S})\right)\boldsymbol{Y}.\end{aligned}$$

所以有

$$\hat{\boldsymbol{\alpha}}_k=(\hat{\alpha}_k(T_1),\hat{\alpha}_k(T_2),\cdots,\hat{\alpha}_k(T_n))^{\mathrm{T}}=\boldsymbol{B}_k\boldsymbol{Y},$$

其中

$$\boldsymbol{B}_k=\begin{pmatrix}\boldsymbol{e}_{k+1,k+1}^{\mathrm{T}}\{\boldsymbol{D}_{T_1}^{\mathrm{T}}\boldsymbol{W}_{T_1}\boldsymbol{D}_{T_1}\}^{-1}\boldsymbol{D}_{T_1}^{\mathrm{T}}\boldsymbol{W}_{T_1}\\\boldsymbol{e}_{k+1,k+1}^{\mathrm{T}}\{\boldsymbol{D}_{T_2}^{\mathrm{T}}\boldsymbol{W}_{T_2}\boldsymbol{D}_{T_2}\}^{-1}\boldsymbol{D}_{T_2}^{\mathrm{T}}\boldsymbol{W}_{T_2}\\\vdots\\\boldsymbol{e}_{k+1,k+1}^{\mathrm{T}}\{\boldsymbol{D}_{T_n}^{\mathrm{T}}\boldsymbol{W}_{T_n}\boldsymbol{D}_{T_n}\}^{-1}\boldsymbol{D}_{T_n}^{\mathrm{T}}\boldsymbol{W}_{T_n}\end{pmatrix}$$
$$\left(\boldsymbol{I}-\boldsymbol{X}[\boldsymbol{X}^{\mathrm{T}}(\boldsymbol{I}-\boldsymbol{S})^{\mathrm{T}}(\boldsymbol{I}-\boldsymbol{S})\boldsymbol{X}]^{-1}\boldsymbol{X}^{\mathrm{T}}(\boldsymbol{I}-\boldsymbol{S})^{\mathrm{T}}(\boldsymbol{I}-\boldsymbol{S})\right).$$

类似于 Gijbels 和 Rousson (2001) 与 Mei 等 (2003), 将检验问题

$$H_0:\quad f(T)=\theta_0+\theta_1T+\cdots+\theta_kT^k$$

转化为检验 $H_0: E\alpha_k(\cdot) = \theta_k$. 以 $\hat{\alpha}_k(T_i)$ 的样本方差为基础来构造检验统计量. 定义

$$
\begin{aligned}
V_n^2(k) &= \frac{1}{n-1}\sum_{i=1}^{n}\left(\hat{\alpha}_k(T_i) - \frac{1}{n}\sum_{j=1}^{n}\hat{\alpha}_k(T_j)\right)^2 \\
&= \frac{1}{n-1}\hat{\boldsymbol{\alpha}}_k^{\mathrm{T}}\left(\boldsymbol{I} - \frac{1}{n}\boldsymbol{J}\right)\hat{\boldsymbol{\alpha}}_k \\
&= \frac{1}{n-1}\boldsymbol{Y}^{\mathrm{T}}\boldsymbol{B}_k^{\mathrm{T}}\left(\boldsymbol{I} - \frac{1}{n}\boldsymbol{J}\right)\boldsymbol{B}_k\boldsymbol{Y}.
\end{aligned} \tag{9.18}
$$

当原假设成立时, 根据定理 9.3, 有

$$E\hat{\boldsymbol{\alpha}}_k = \theta_k \mathbf{1}_n,$$

又因为

$$\mathbf{1}_n^{\mathrm{T}}\left(\boldsymbol{I} - \frac{1}{n}\boldsymbol{J}\right) = \mathbf{0}, \quad \left(\boldsymbol{I} - \frac{1}{n}\boldsymbol{J}\right)\mathbf{1}_n = \mathbf{0},$$

所以可得

$$
\begin{aligned}
(n-1)V_n^2(k) &= \hat{\boldsymbol{\alpha}}_k^{\mathrm{T}}\left(\boldsymbol{I} - \frac{1}{n}\boldsymbol{J}\right)\hat{\boldsymbol{\alpha}}_k \\
&= (\hat{\boldsymbol{\alpha}}_k - E\hat{\boldsymbol{\alpha}}_k)^{\mathrm{T}}\left(\boldsymbol{I} - \frac{1}{n}\boldsymbol{J}\right)(\hat{\boldsymbol{\alpha}}_k - E\hat{\boldsymbol{\alpha}}_k) \\
&= \boldsymbol{\varepsilon}^{\mathrm{T}}\boldsymbol{B}_k^{\mathrm{T}}\left(\boldsymbol{I} - \frac{1}{n}\boldsymbol{J}\right)\boldsymbol{B}_k\boldsymbol{\varepsilon}.
\end{aligned}
$$

由于 $V_n^2(k)$ 的分布与误差方差 σ^2 有关, 所以我们必须给出其估计值. 在部分线性模型的研究中, σ^2 的常用估计是 RSS_1/n. 虽然该估计具有良好的性质, 但是其估计中包含有窗宽, 窗宽的选择对检验的结果影响往往很大. 所以我们建议使用下面的差分方法估计 σ^2.

对于部分线性模型 (9.2), 假设 $0 \leqslant T_1 < T_2 < \cdots < T_n \leqslant 1$, 利用一阶差分得如下的线性回归模型

$$
\begin{aligned}
Y_{i+1} - Y_i &= (f(T_{i+1}) - f(T_i)) + (\boldsymbol{X}_{i+1} - \boldsymbol{X}_i)^{\mathrm{T}}\boldsymbol{\beta} + \varepsilon_{i+1} - \varepsilon_i \\
&\approx (\boldsymbol{X}_{i+1} - \boldsymbol{X}_i)^{\mathrm{T}}\boldsymbol{\beta} + \varepsilon_{i+1} - \varepsilon_i,
\end{aligned}
$$

基于一般线性回归模型的理论, 构造 σ^2 的差分估计为

$$\hat{\sigma}^2 = \frac{1}{2(n-1)}\sum_{i=1}^{n-1}\left(Y_{i+1} - Y_i - (\boldsymbol{X}_{i+1} - \boldsymbol{X}_i)^{\mathrm{T}}\hat{\boldsymbol{\beta}}_{\mathrm{dif}}\right)^2,$$

其中 $\hat{\boldsymbol{\beta}}_{\text{dif}}$ 为 $\boldsymbol{\beta}$ 的最小二乘估计. 令

$$\boldsymbol{D} = \begin{pmatrix} -1 & 1 & 0 & \cdots & 0 & 0 \\ 0 & -1 & 1 & \cdots & 0 & 0 \\ \vdots & \vdots & \vdots & & \vdots & \vdots \\ 0 & 0 & 0 & \cdots & -1 & 1 \end{pmatrix}_{(n-1)\times n},$$

则有 $\hat{\sigma}^2$ 的矩阵表示为

$$\hat{\sigma}^2 = \frac{1}{2(n-1)}\boldsymbol{Y}^{\mathrm{T}}\boldsymbol{D}^{\mathrm{T}}(\boldsymbol{I}-\boldsymbol{D}\boldsymbol{X}(\boldsymbol{X}^{\mathrm{T}}\boldsymbol{D}^{\mathrm{T}}\boldsymbol{D}\boldsymbol{X})^{-1}\boldsymbol{X}^{\mathrm{T}}\boldsymbol{D}^{\mathrm{T}})\boldsymbol{D}\boldsymbol{Y}. \tag{9.19}$$

在备择假设与原假设情况下, 近似地有

$$2(n-1)\hat{\sigma}^2 = \boldsymbol{\varepsilon}^{\mathrm{T}}\boldsymbol{D}^{\mathrm{T}}(\boldsymbol{I}-\boldsymbol{D}\boldsymbol{X}(\boldsymbol{X}^{\mathrm{T}}\boldsymbol{D}^{\mathrm{T}}\boldsymbol{D}\boldsymbol{X})^{-1}\boldsymbol{X}^{\mathrm{T}}\boldsymbol{D}^{\mathrm{T}})\boldsymbol{D}\boldsymbol{\varepsilon}.$$

基于式 (9.18) 和式 (9.19), 构造检验统计量如下

$$\mathrm{T}_2 = \frac{\boldsymbol{Y}^{\mathrm{T}}\boldsymbol{B}_k^{\mathrm{T}}(\boldsymbol{I}-\frac{1}{n}\boldsymbol{J})\boldsymbol{B}_k\boldsymbol{Y}}{\boldsymbol{Y}^{\mathrm{T}}\boldsymbol{D}^{\mathrm{T}}(\boldsymbol{I}-\boldsymbol{D}\boldsymbol{X}(\boldsymbol{X}^{\mathrm{T}}\boldsymbol{D}^{\mathrm{T}}\boldsymbol{D}\boldsymbol{X})^{-1}\boldsymbol{X}^{\mathrm{T}}\boldsymbol{D}^{\mathrm{T}})\boldsymbol{D}\boldsymbol{Y}},$$

则当原假设成立时, 有

$$T_2 = \frac{\boldsymbol{\varepsilon}^{\mathrm{T}}\boldsymbol{B}_k^{\mathrm{T}}(\boldsymbol{I}-\frac{1}{n}\boldsymbol{J})\boldsymbol{B}_k\boldsymbol{\varepsilon}}{\boldsymbol{\varepsilon}^{\mathrm{T}}\boldsymbol{D}^{\mathrm{T}}(\boldsymbol{I}-\boldsymbol{D}\boldsymbol{X}(\boldsymbol{X}^{\mathrm{T}}\boldsymbol{D}^{\mathrm{T}}\boldsymbol{D}\boldsymbol{X})^{-1}\boldsymbol{X}^{\mathrm{T}}\boldsymbol{D}^{\mathrm{T}})\boldsymbol{D}\boldsymbol{\varepsilon}}.$$

显然对于该类型的检验统计量, 如果模型误差服从正态分布, 我们可以利用推论 1.1~ 推论 1.3 来求得检验 p 值.

9.4 部分线性模型的 profile 局部加权最小二乘估计

实际数据分析中, 若我们主观的将因变量与自变量之间的关系设定为参数关系, 那么很多情况下对问题的分析要冒较大的风险, 甚至得出不合实际的结论. 但如果置经验于不顾而利用非参数回归模型, 那么对回归函数假设放宽的代价是拟合相应的非参数回归模型的计算量的增加以及当某个参数模型适合于所分析的数据时, 非参数模型便没有参数模型那么有效. 针对这一两难问题, 很多作者提出了将参数回归与非参数相结合的估计方法, 详细内容可参见 Hjort 和 Jones (1996), Fan 和 Ullah (1999) 和 Gozalo 和 Linton (2000), 其中应用最为广泛的是 Gozalo 和 Linton (2000) 提出的局部非线性最小二乘法. 下面基于非参数回归模型 (1.4) 简要叙述该方法.

首先根据经验和数据信息将因变量 Y 与自变量 X 的关系设定为如下的参数形式

$$m(x) = m(x, \boldsymbol{\theta}),$$

对于其中未知参数 $\boldsymbol{\theta}$ 的估计, 我们不采用传统的 (非线性) 最小二乘估计方法, 而是利用如下的局部 (非线性) 最小二乘法估计, 即通过极小化下式

$$\sum_{i=1}^{n}\{y_i - m(x_i, \boldsymbol{\theta})\}^2 K_h(x_i - x)$$

得到. 记对应的估计值为 $\hat{\boldsymbol{\theta}}(x)$, 从而得到 $m(x)$ 的估计值为 $\hat{m}(x, \hat{\boldsymbol{\theta}}(x))$. 值得注意的是, 如果 $m(x, \boldsymbol{\theta}) = \theta$, 那么 $\hat{m}(x) = \hat{\theta}(x)$ 就是 N-W 核估计. 如果 $m(x, \boldsymbol{\theta}) = \theta_0 + \theta_1 x + \cdots + \theta_p x^k$, 其中 $p \geqslant 1$, 那么 $\hat{m}(x) = \hat{\theta}_0(x) + \hat{\theta}_1(x)x + \cdots + \hat{\theta}_p(x)x^p$ 就是阶数为 p 的局部多项式估计拟合值. 另外, Gozalo 和 Linton (2000) 证明了该估计的渐近方差与参数模型无关, 并且其渐近偏差为

$$\text{Bias } \hat{m}(x) = \text{Bias } \hat{m}(x, \hat{\boldsymbol{\theta}}(x)) = \frac{1}{2}\mu_2 h^2 (m^{(2)}(x) - m^{(2)}(x, \boldsymbol{\theta}(x))).$$

显然其偏差与 x 的密度 f 无关, 从而该估计为自适应的. 其偏差决定于参数模型 $m(x, \boldsymbol{\theta})$ 与非参数模型 $m(x)$ 之间的偏离程度. 如果对于所有的 x,$m(x) = m(x, \boldsymbol{\theta})$ 都成立, 那么 $\hat{m}(x, \hat{\boldsymbol{\theta}}(x))$ 为无偏估计, 然而 (p 阶) 局部多项式估计只有当均值函数为一 (小于等于 p 阶) 多项式时才为无偏估计. 并且如果 $|m^{(2)}(x) - m^{(2)}(x, \boldsymbol{\theta}(x))| \leqslant |m^{(2)}(x)|$, 那么 $\hat{m}(x, \hat{\boldsymbol{\theta}}(x))$ 的偏差要小于核估计以及局部线性估计的偏差.

下面将该方法引入到部分线性模型的估计中, 并且在该估计方法的基础上利用部分线性模型检验一般的线性关系.

对于部分线性模型 (9.2)

$$Y_i = f(T_i) + \boldsymbol{X}_i^{\mathrm{T}} \boldsymbol{\beta} + \varepsilon_i.$$

前面介绍了 profile 最小二乘估计方法, 下面给出 profile 局部加权最小二乘估计, 该方法在半参数模型的估计中利用了参数模型的信息. 对于非参数部分 $f(\cdot)$, 首先根据经验或者数据本身提供的信息, 将其设定为如下的参数模型

$$f(T_i) = g_1(T_i)\theta_1 + g_2(T_i)\theta_2 + \cdots + g_q(T_i)\theta_q = \boldsymbol{g}^{\mathrm{T}}(T_i)\boldsymbol{\theta}. \tag{9.20}$$

首先假定模型 (9.2) 中 $(\beta_1, \beta_2, \cdots, \beta_p)^{\mathrm{T}}$ 已知 (给定), 则模型 (9.2) 转化为如下形式的非参数回归模型

$$Y_i^* = f(T_i) + \varepsilon_i, \quad i = 1, 2, \cdots, n, \tag{9.21}$$

其中 $Y_i^* = Y_i - \boldsymbol{X}_i^{\mathrm{T}}\boldsymbol{\beta}$. 对于非参数回归模型 (9.21), 结合 $f(\cdot)$ 的参数设定模型 (9.20), 利用前面介绍的 Gozalo 和 Linton (2000) 的局部加权最小二乘法可得 $f(T_i)$ 的估计值为

$$\hat{f}(T_i) = \boldsymbol{g}^{\mathrm{T}}(T_i)\hat{\boldsymbol{\theta}}(T_i),$$

其中 $\hat{\boldsymbol{\theta}}(T_i)$ 为针对 $\boldsymbol{\theta}(T_i)$ 使

$$\sum_{j=1}^{n}\{Y_j^* - \boldsymbol{g}^{\mathrm{T}}(T_j)\boldsymbol{\theta}(T_i)\}^2 K_h(T_j - T_i)$$

达到最小得到.

令

$$\boldsymbol{G} = \begin{pmatrix} \boldsymbol{g}^{\mathrm{T}}T_1 \\ \boldsymbol{g}^{\mathrm{T}}T_2 \\ \vdots \\ \boldsymbol{g}^{\mathrm{T}}T_n \end{pmatrix} = \begin{pmatrix} g_1(T_1) & \cdots & g_q(T_1) \\ g_1(T_2) & \cdots & g_q(T_2) \\ \vdots & & \vdots \\ g_1(T_q) & \cdots & g_q(T_q) \end{pmatrix}, \quad \boldsymbol{f} = \begin{pmatrix} f(T_1) \\ f(T_2) \\ \vdots \\ f(T_n) \end{pmatrix}, \quad \boldsymbol{\varepsilon} = \begin{pmatrix} \varepsilon_1 \\ \varepsilon_2 \\ \vdots \\ \varepsilon_n \end{pmatrix},$$

以及 $\boldsymbol{Y} = (y_1, y_2, \cdots, y_n)^{\mathrm{T}}$, $\boldsymbol{W}_{T_i} = \mathrm{diag}(K_h(T_1 - T_i), K_h(T_2 - T_i), \cdots, K_h(T_n - T_i))$, 则有

$$\hat{\boldsymbol{\theta}}(T_i) = \left(\boldsymbol{G}^{\mathrm{T}}\boldsymbol{W}_{T_i}\boldsymbol{G}\right)\boldsymbol{G}^{\mathrm{T}}\boldsymbol{W}_{T_i}(\boldsymbol{Y} - \boldsymbol{X}\boldsymbol{\beta}),$$

从而有

$$\hat{f}(T_i) = \boldsymbol{g}^{\mathrm{T}}(T_i)\hat{\boldsymbol{\theta}}(T_i) = \boldsymbol{g}^{\mathrm{T}}(T_i)\left(\boldsymbol{G}^{\mathrm{T}}\boldsymbol{W}_{T_i}\boldsymbol{G}\right)\boldsymbol{G}^{\mathrm{T}}\boldsymbol{W}_{T_i}(\boldsymbol{Y} - \boldsymbol{X}\boldsymbol{\beta}),$$

则 $\boldsymbol{f}$ 的估计为

$$\hat{\boldsymbol{f}} = \boldsymbol{S}(\boldsymbol{Y} - \boldsymbol{X}\boldsymbol{\beta}),$$

其中

$$\boldsymbol{S} = \begin{pmatrix} \boldsymbol{g}^{\mathrm{T}}(T_1)\left(\boldsymbol{G}^{\mathrm{T}}\boldsymbol{W}_{T_1}\boldsymbol{G}\right)\boldsymbol{G}^{\mathrm{T}}\boldsymbol{W}_{T_1} \\ \boldsymbol{g}^{\mathrm{T}}(T_2)\left(\boldsymbol{G}^{\mathrm{T}}\boldsymbol{W}_{T_2}\boldsymbol{G}\right)\boldsymbol{G}^{\mathrm{T}}\boldsymbol{W}_{T_2} \\ \vdots \\ \boldsymbol{g}^{\mathrm{T}}(T_n)\left(\boldsymbol{G}^{\mathrm{T}}\boldsymbol{W}_{T_n}\boldsymbol{G}\right)\boldsymbol{G}^{\mathrm{T}}\boldsymbol{W}_{T_n} \end{pmatrix}.$$

将上面 $\boldsymbol{f}$ 的估计代入原模型 (9.2) 整理可得如下的线性回归模型

$$\overline{\boldsymbol{Y}} = \overline{\boldsymbol{X}}\boldsymbol{\beta} + \boldsymbol{\varepsilon}, \tag{9.22}$$

其中

$$\overline{\boldsymbol{Y}} = (\boldsymbol{I} - \boldsymbol{S})\boldsymbol{Y}, \quad \overline{\boldsymbol{X}} = (\boldsymbol{I} - \boldsymbol{S})\boldsymbol{X}.$$

利用最小二乘法估计上面的模型, 则得 $\boldsymbol{\beta}$ 的 profile 局部加权最小二乘估计为

$$\hat{\boldsymbol{\beta}} = [\boldsymbol{X}^{\mathrm{T}}(\boldsymbol{I}-\boldsymbol{S})^{\mathrm{T}}(\boldsymbol{I}-\boldsymbol{S})\boldsymbol{X}]^{-1}\boldsymbol{X}^{\mathrm{T}}(\boldsymbol{I}-\boldsymbol{S})^{\mathrm{T}}(\boldsymbol{I}-\boldsymbol{S})\boldsymbol{Y}, \tag{9.23}$$

得 $\boldsymbol{f}$ 的估计为

$$\hat{\boldsymbol{f}} = \boldsymbol{S}(\boldsymbol{Y}-\boldsymbol{X}\hat{\boldsymbol{\beta}}). \tag{9.24}$$

从而 $\boldsymbol{Y}$ 的拟合值为

$$\hat{\boldsymbol{Y}} = \hat{\boldsymbol{f}} + \boldsymbol{X}\hat{\boldsymbol{\beta}} = \boldsymbol{L}_1\boldsymbol{Y}, \tag{9.25}$$

残差平方和为

$$\mathrm{RSS} = \boldsymbol{Y}^{\mathrm{T}}(\boldsymbol{I}-\boldsymbol{L})^{\mathrm{T}}(\boldsymbol{I}-\boldsymbol{L})\boldsymbol{Y}, \tag{9.26}$$

其中

$$\boldsymbol{L} = \boldsymbol{S} + (\boldsymbol{I}-\boldsymbol{S})\boldsymbol{Z}[\boldsymbol{Z}^{\mathrm{T}}(\boldsymbol{I}-\boldsymbol{S})^{\mathrm{T}}(\boldsymbol{I}-\boldsymbol{S})\boldsymbol{Z}]^{-1}\boldsymbol{Z}^{\mathrm{T}}(\boldsymbol{I}-\boldsymbol{S})^{\mathrm{T}}(\boldsymbol{I}-\boldsymbol{S}).$$

关于该估计, 有如下的结论.

定理 9.5 如果原假设 H_0 成立, 对于 profile 局部线性最小二乘估计有

$$E\hat{\boldsymbol{\beta}} = \boldsymbol{\beta}; \quad E\hat{f}(\mathrm{T}_i) = f(T_i) = \boldsymbol{g}^{\mathrm{T}}(T_i)\boldsymbol{\theta}.$$

对于残差平方和 RSS, 有 $\mathrm{RSS} = \boldsymbol{\varepsilon}^{\mathrm{T}}(\boldsymbol{I}-\boldsymbol{L})^{\mathrm{T}}(\boldsymbol{I}-\boldsymbol{L})\boldsymbol{\varepsilon}$.

前面的 5.2 节和 5.3 节讨论了利用部分线性模型检验线性关系, 构造了广义似然比检验统计量和一种新的检验统计量. 然而值得注意的是, 这两种方法主要是检验非参数部分是否为某一特定阶数的多项式, 对于这类特殊的线性关系成立时, 相应的局部多项式估计为无偏估计, 很多检验统计量的构造也是基于这一性质. 然而如果要检验的线性关系不是某一特定阶数的多项式时, 前面检验统计量的研究就会变的相对复杂. 为了克服这一困难, 结合前面提出的 profile 局部加权最小二乘估计来构造检验统计量进行一般线性关系的检验. 不失一般性, 记原假设为

$$H_0^*: f(T) = g_1(T)\theta_1 + g_2(T)\theta_2 + \cdots + g_q(T)\theta_q = \boldsymbol{g}^{\mathrm{T}}(T)\boldsymbol{\theta}, \quad \forall T.$$

显然上面的检验问题包括 9.2 节研究的检验问题 (9.3). 下面给出检验统计量的构造过程. 首先当原假设成立时有

$$f(T_i) = g_1(T_i)\theta_1 + g_2(T_i)\theta_2 + \cdots + g_p(T_i)\theta_q = \boldsymbol{g}^{\mathrm{T}}(T_i)\boldsymbol{\theta}. \tag{9.27}$$

那么部分线性模型 (9.2) 变为如下的线性模型

$$Y_i = \boldsymbol{g}^{\mathrm{T}}(T_i)\boldsymbol{\theta} + \boldsymbol{x}_i^{\mathrm{T}}\boldsymbol{\beta} + \varepsilon_i.$$

记该线性模型基于最小二乘估计方法得到的残差平方和 RSS_0. 备择假设下, 与第二节不同的是, 对于部分线性模型的估计不再利用 profile 局部多项式最小二乘估计, 而是结合 $f(\cdot)$ 的参数设定形式 (9.27) 利用前面提出的 profile 局部加权最小二乘估计方法, 记相应的残差平方和为 RSS_1. 构造检验统计量

$$T_3=\frac{\mathrm{RSS}_0-\mathrm{RSS}_1}{\mathrm{RSS}_1}.$$

显然, 基于最小二乘估计的性质和定理 9.5, 当原假设成立时, 上 T_N 可化为如下的形式

$$T_3=\frac{\boldsymbol{\varepsilon}^{\mathrm{T}}\left\{\boldsymbol{I}-\boldsymbol{Z}(\boldsymbol{Z}^{\mathrm{T}}\boldsymbol{Z})^{-1}\boldsymbol{Z}^{\mathrm{T}}-(\boldsymbol{I}-\boldsymbol{L})^{\mathrm{T}}(\boldsymbol{I}-\boldsymbol{L})\right\}\boldsymbol{\varepsilon}}{\boldsymbol{\varepsilon}^{\mathrm{T}}(\boldsymbol{I}-\boldsymbol{L})^{\mathrm{T}}(\boldsymbol{I}-\boldsymbol{L})\boldsymbol{\varepsilon}},$$

其中

$$\boldsymbol{Z}=\begin{pmatrix} g_1(T_1) & \cdots & g_q(T_1) & x_{11} & \cdots & x_{1p}\\ g_1(T_2) & \cdots & g_q(T_2) & x_{21} & \cdots & x_{2p}\\ \vdots & & \vdots & \vdots & & \vdots\\ g_1(T_n) & \cdots & g_q(T_n) & x_{n1} & \cdots & x_{np} \end{pmatrix}.$$

显然对于该类型的检验统计量, 如果模型误差服从正态分布, 可以利用结论 1.1~ 结论 1.3 来求得检验 p 值.

9.5 定理的证明

下面给出定理 9.1~ 定理 9.5 的证明, 定理 9.3 是结论 1.1~ 结论 1.3 的直接结论, 故其证明过程在此省略.

在给出定理 9.1 的证明过程之前, 给出下面所需要的条件.

(A.1) 随机变量 T 具有有界支撑 $\boldsymbol{\Omega}$, 并且其边际密度函数 $g(\cdot)$ 在其支撑上满足 Lipschitz 连续, 且不为 0.

(A.2) $f(\cdot)$ 具有 $k+1$ 阶连续导函数.

(A.3) 函数 $K(\cdot)$ 为对称密度函数, 具有紧支撑, 并且满足 Lipschitz 连续.

定理 9.1 的证明 假设 $\boldsymbol{\beta}$ 是参数的真实值, 令 $Y^*=Y-\boldsymbol{X}^{\mathrm{T}}\boldsymbol{\beta}$, 从而将部分线性模型 (9.2) 转变为如下的标准非参数回归模型

$$Y_i^*=f(T_i)+\varepsilon_i.$$

对于检验问题 H_0^N, 基于上面的非参数回归模型可构造如下的广义似然比检验统计量

$$T_N^*=\frac{n}{2}\log\frac{\mathrm{RSS}_0^*}{\mathrm{RSS}_1^*},$$

其中 $\mathrm{RSS}_0^* = \sum_{i=1}^n \{Y_i^* - \overline{\alpha}_1 - \overline{\alpha}_2 T_i\}^2, \mathrm{RSS}_1^* = \sum_{i=1}^n \left\{Y_i^* - \hat{f}(T_i)\right\}^2$.

根据 Fan 等 (2001), 有

$$r_k T_N^* \sim \chi^2_{\delta_n}.$$

因此只需证明下面两个结论成立

$$\text{(a)} \quad n^{-1}\{\mathrm{RSS}_0 - \mathrm{RSS}_0^*\} = o_p(1),$$

与

$$\text{(b)} \quad n^{-1}\{\mathrm{RSS}_1 - \mathrm{RSS}_1^*\} = o_p(1).$$

由线性回归模型的经典理论有

$$\mathrm{RSS}_0^* = \sigma^2(n-p)\{1+o_p(1)\}, \quad \mathrm{RSS}_0 = \sigma^2(n-p-2)\{1+o_p(1)\}.$$

基于上面的两个结论, 直接可推得 (a) 成立. 根据

$$\mathrm{RSS}_1 - \mathrm{RSS}_1^* = (\hat{\boldsymbol{\beta}} - \boldsymbol{\beta})^{\mathrm{T}} \overline{\boldsymbol{X}}^{\mathrm{T}} \overline{\boldsymbol{X}} (\hat{\boldsymbol{\beta}} - \boldsymbol{\beta}) = o_p(n),$$

从而结论 (b) 成立.

定理 9.2 和 9.4 的证明　下面记

$$\boldsymbol{T} = \begin{pmatrix} 1 & T_1 & \cdots & T_1^k \\ 1 & T_2 & \cdots & T_2^k \\ \vdots & \vdots & & \vdots \\ 1 & T_n & \cdots & T_n^k \end{pmatrix}.$$

显然原假设成立有

$$\boldsymbol{Y} = \boldsymbol{T}\boldsymbol{\theta} + \boldsymbol{X}\boldsymbol{\beta}.$$

首先对 $\boldsymbol{T}\boldsymbol{\theta}$ 进行分解可得

$$\begin{aligned} \boldsymbol{T}\boldsymbol{\theta} &= \begin{pmatrix} \theta_0 + \theta_1 T_1 + \cdots + \theta_k T_1^k \\ \theta_0 + \theta_1 T_2 + \cdots + \theta_k T_2^k \\ \vdots \\ \theta_0 + \theta_1 T_n + \cdots + \theta_k T_n^k \end{pmatrix} \\ &= \begin{pmatrix} 1 & T_1 - T_i & \cdots & (T_1 - T_i)^k \\ 1 & T_2 - T_i & \cdots & (T_2 - T_i)^k \\ \vdots & \vdots & & \vdots \\ 1 & T_n - T_i & \cdots & (T_n - T_i)^k \end{pmatrix} \begin{pmatrix} \phi_0 \\ \phi_1 \\ \vdots \\ \phi_k \end{pmatrix} \end{aligned}$$

$$= \boldsymbol{D}_{T_i}\boldsymbol{\phi}.$$

显然对于 $\boldsymbol{\phi}$ 有

$$\phi_0 = \theta_0 + \theta_1 T_i + \cdots + \theta_k T_i^k, \quad \phi_k = \theta_k.$$

从而可得

$$\boldsymbol{e}_{1,k+1}^{\mathrm{T}}\{\boldsymbol{D}_{T_i}^{\mathrm{T}}\boldsymbol{W}_{T_i}\boldsymbol{D}_{T_i}\}^{-1}\boldsymbol{D}_{T_i}^{\mathrm{T}}\boldsymbol{W}_{T_i}\boldsymbol{T\theta} = \phi_0 = \theta_0 + \theta_1 T_i + \cdots + \theta_k T_i^k,$$

与

$$\boldsymbol{e}_{k+1,k+1}^{\mathrm{T}}\{\boldsymbol{D}_{T_i}^{\mathrm{T}}\boldsymbol{W}_{T_i}\boldsymbol{D}_{T_i}\}^{-1}\boldsymbol{D}_{T_i}^{\mathrm{T}}\boldsymbol{W}_{T_i}\boldsymbol{T\theta} = \phi_k = \theta_k.$$

基于这个结论, 有

$$\begin{aligned}\boldsymbol{ST\theta} &= \begin{pmatrix} \boldsymbol{e}_{1,k+1}^{\mathrm{T}}\{\boldsymbol{D}_{T_1}^{\mathrm{T}}\boldsymbol{W}_{T_1}\boldsymbol{D}_{T_1}\}^{-1}\boldsymbol{D}_{T_1}^{\mathrm{T}}\boldsymbol{W}_{T_1}\boldsymbol{T\alpha} \\ \boldsymbol{e}_{1,k+1}^{\mathrm{T}}\{\boldsymbol{D}_{T_2}^{\mathrm{T}}\boldsymbol{W}_{T_2}\boldsymbol{D}_{T_2}\}^{-1}\boldsymbol{D}_{T_2}^{\mathrm{T}}\boldsymbol{W}_{T_2}\boldsymbol{T\alpha} \\ \vdots \\ \boldsymbol{e}_{1,k+1}^{\mathrm{T}}\{\boldsymbol{D}_{T_n}^{\mathrm{T}}\boldsymbol{W}_{T_n}\boldsymbol{D}_{T_n}\}^{-1}\boldsymbol{D}_{T_n}^{\mathrm{T}}\boldsymbol{W}_{T_n}\boldsymbol{T\alpha} \end{pmatrix} \\ &= \begin{pmatrix} \theta_0 + \theta_1 T_1 + \cdots + \theta_k T_1^k \\ \theta_0 + \theta_1 T_2 + \cdots + \theta_k T_2^k \\ \vdots \\ \theta_0 + \theta_1 T_n + \cdots + \theta_k T_n^k \end{pmatrix} = \boldsymbol{T\theta}.\end{aligned}$$

下面来证明 $\hat{\boldsymbol{\beta}}$ 以及 $\hat{\boldsymbol{f}}$, $\hat{\boldsymbol{\alpha}}_k$ 在原假设成立时的无偏性.

$$E\hat{\boldsymbol{\beta}} = \boldsymbol{\beta} + [\boldsymbol{X}^{\mathrm{T}}(\boldsymbol{I}-\boldsymbol{S})^{\mathrm{T}}(\boldsymbol{I}-\boldsymbol{S})\boldsymbol{X}]^{-1}\boldsymbol{X}^{\mathrm{T}}(\boldsymbol{I}-\boldsymbol{S})^{\mathrm{T}}(\boldsymbol{I}-\boldsymbol{S})\mathbf{T}\boldsymbol{\theta} = \boldsymbol{\beta}$$

与

$$E\hat{\boldsymbol{f}} = E\boldsymbol{S}(\boldsymbol{Y} - \boldsymbol{X}\hat{\boldsymbol{\theta}}) = E\boldsymbol{ST\alpha} = \boldsymbol{T\theta} = \boldsymbol{f},$$

另外

$$E\hat{\boldsymbol{\alpha}}_k = E\boldsymbol{B}_k\boldsymbol{Y} = E\boldsymbol{B}_k\boldsymbol{T\theta} = \mathbf{1}_n\theta_k.$$

由上面的结论, 可得

$$E(\hat{\boldsymbol{Y}}) = E(\hat{\boldsymbol{f}} + \boldsymbol{X}\hat{\boldsymbol{\beta}}) = E(\boldsymbol{Y}).$$

令 $\hat{\boldsymbol{\varepsilon}} = (\hat{\varepsilon}_1, \hat{\varepsilon}_2, \cdots, \hat{\varepsilon}_n)^{\mathrm{T}}$ 为模型 (9.2) 基于 profile 最小二乘估计的拟合残差. 显然有

$$E(\hat{\boldsymbol{\varepsilon}}) = E(\boldsymbol{Y}) - E(\hat{\boldsymbol{Y}}) = \mathbf{0},$$

则 RSS_1 可表示为

$$\begin{aligned}\mathrm{RSS}_1 &= \hat{\boldsymbol{\varepsilon}}^{\mathrm{T}}\hat{\boldsymbol{\varepsilon}} = [\hat{\boldsymbol{\varepsilon}} - E(\hat{\boldsymbol{\varepsilon}})]^{\mathrm{T}}[\hat{\boldsymbol{\varepsilon}} - E(\hat{\boldsymbol{\varepsilon}})] \\ &= [\boldsymbol{Y} - E(\hat{\boldsymbol{Y}})]^{\mathrm{T}}(\boldsymbol{I} - \boldsymbol{L})^{\mathrm{T}}(\boldsymbol{I} - \boldsymbol{L})[\boldsymbol{Y} - E(\hat{\boldsymbol{Y}})] \\ &= \boldsymbol{\varepsilon}^{\mathrm{T}}(\boldsymbol{I} - \boldsymbol{L})^{\mathrm{T}}(\boldsymbol{I} - \boldsymbol{L})\boldsymbol{\varepsilon}.\end{aligned}$$

定理 9.5 的证明　基于 $\boldsymbol{SG\theta} = \boldsymbol{G\theta}$, 估计的无偏性立得, RSS 的证明类似定理 9.2, 故在此省略.

第10章　非参数协方差分析模型的研究

近年来, 非参数协方差分析模型即非参数回归曲线的比较得到了人们的关注, 已经有多种方法提出. 正如方差分析模型与协方差分析模型作为线性回归模型的特例从而可以直接应用关于线性回归模型的估计与检验的一些结论, 本章将基于非参数协方差分析模型是一类特殊的变系数模型这一事实进行几个方面的研究. 首先对于两条曲线的比较提出了虚拟变量检验法, 该方法将关于两条曲线的相等以及平行的检验转化为变系数模型中关于模型系数的检验问题. 其次基于残差分析的思想提出了两种检验方法. 对于特殊的一类非参数协方差分析模型, 即各条曲线对应的自变量相同, 我们给出了总体均值曲线与个体差异曲线的估计以及基于该估计构造了检验统计量, 并对几种检验方法的关系作了研究, 最后在变系数模型的分析框架下讨论了一般的非参数协方差分析模型的广义似然比检验统计量.

10.1　引　言

不同数据集上因变量与自变量之间回归关系的比较是应用回归分析中非常重要的一个问题, 比如, 生物医学试验中常见的控制组与对照组效果的比较, 经济学中利用收入函数检验性别或种族歧视的存在以及经济结构稳定性的检验等. 方差分析模型与协方差分析模型作为处理此类问题的有效方法得到了广泛的研究与应用. 由回归分析的经典理论可知, 回归函数的不正确设定会导致推断的严重错误, 从而会使得最后的结论严重偏离实际情况, 为了克服参数模型的这一缺陷, 非参数协方差分析模型的研究近年来引起了人们的重视. 一般情况下我们记模型为下面的形式

$$Y_{ij} = f_i(x_{ij}) + \varepsilon_{ij}, \quad i = 1, 2, \cdots, p, \quad j = 1, 2, \cdots, n_i, \tag{10.1}$$

其中 Y_{ij}, x_{ij} 为因变量与自变量观测值, ε_{ij} 为模型误差, f_i 是对应于第 i 组观测数据的未知回归函数. 需要研究的就是比较这 p 条回归曲线的异同, 即为如下的检验问题

$$H_0: \quad f_1 = f_2 = \cdots = f_p \qquad \text{VS} \qquad H_1: \quad f_i \neq f_j, \quad \text{对某些 } i,\ j \in \{1, \cdots, p\}. \tag{10.2}$$

对于上面的检验问题, 已经有多种方法提出, 其中 King 等 (1991) 及 Hall 和 Hart (1990) 讨论了自变量相同的情况下两条非参数回归曲线的比较. Young 和

Bowman (1995) 利用方差分析的思想对多条非参数回归曲线是否相等和平行进行检验. 而 Yatchew (1999) 和 Dette 和 Neumeyer (2001) 则基于比较方差的办法来解决该问题. 其他还有基于经验过程、秩统计量、Wald 方法构造的检验统计量.

正如方差分析模型与协方差分析模型是特殊的线性模型一样, 注意到非参数协方差分析模型是一类特殊的变系数模型, 本质上是一种条件方差分析模型. 与前面提到的方法不同, 本书将基于这一特点将回归曲线的比较问题转化为变系数模型的检验问题, 从而可以直接应用有关变系数模型检验的一些结果, 或者构造新的检验统计量.

10.2 两条回归曲线比较的虚拟变量法

实际问题研究中, 经常遇到的是两个数据集上的回归函数的比较, 而且两个回归函数的比较问题的研究是多个回归函数比较问题的研究的基础. 本节重点研究两条回归曲线的比较问题. 考虑如下两条回归曲线

$$Y_{ij} = f_i(x_{ij}) + +\varepsilon_{ij}, \quad i = 1, 2; \quad j = 1, 2, \cdots, n_i,$$

其中 Y_{ij} 为因变量观测值, 而 x_{ij} 为自变量观测值, 不失一般性, 假设 x_{ij} 为一维变量, 随机误差 ε_{ij} 独立同分布, 假设其均值为 0, 方差为 σ^2.

值得说明的是, 如果上面的回归函数 $f_i(\cdot)$ 是线性回归函数, 则回归函数的比较问题转化了两组回归系数的比较问题, Chow (1960) 最早研究了这个问题, 并提出了相应的检验方法 Chow 检验方法. 与 Chow 方法不同的是 Gujarati (1970) 所提出的虚拟变量法.

下面要研究的是这两条未知光滑曲线的比较, 首先研究的是这两条曲线是否相等, 即为如下的检验问题

$$H_{01}: \quad f_1 = f_2, \quad \text{VS} \quad H_{11}: \quad f_1 \neq f_2. \tag{10.3}$$

很多时候, 不仅仅对两条曲线的相等与否感兴趣, 还希望能知道二者是否平行, 即为如下的检验问题

$$H_{02}: \quad f_1 = f_2 + a, \quad a \neq 0 \quad \text{VS} \quad H_{12}: \quad f_1 \neq f_2 + a. \tag{10.4}$$

下面将上面的问题转化为变系数模型中的检验问题, 构造检验统计量.

在备择假设下, 即对两条曲线没有任何约束下, 可将模型 (10.1) 的两个模型合并为一个模型, 即

$$\begin{pmatrix} \boldsymbol{Y}_1 \\ \boldsymbol{Y}_2 \end{pmatrix} = \begin{pmatrix} f_1(\boldsymbol{x}_1) \\ f_2(\boldsymbol{x}_2) \end{pmatrix} + \begin{pmatrix} \varepsilon_1 \\ \varepsilon_2 \end{pmatrix} = g_1 \begin{pmatrix} \boldsymbol{x}_1 \\ \boldsymbol{x}_2 \end{pmatrix} + \begin{pmatrix} \mathbf{0}_{n_1\times 1} \\ g_2(\boldsymbol{x}_2) \end{pmatrix} + \begin{pmatrix} \varepsilon_1 \\ \varepsilon_2 \end{pmatrix},$$

其中, $i=1,2,$

$$\boldsymbol{Y}_i=\begin{pmatrix} y_{i1}\\ y_{i2}\\ \vdots\\ y_{in_i}\end{pmatrix},\quad f_i(\boldsymbol{x}_i)=\begin{pmatrix} f_i(x_{i1})\\ f_i(x_{i2})\\ \vdots\\ f_i(x_{in_i})\end{pmatrix},\quad \boldsymbol{\varepsilon}_i=\begin{pmatrix} \varepsilon_{i1}\\ \varepsilon_{i2}\\ \vdots\\ \varepsilon_{in_i}\end{pmatrix}.$$

由模型的设定可知 $g_1=f_1$, $g_2=f_1-f_2$, 显然上面的模型实质上是一个变系数模型, 可记为

$$y_{ij}=g_1(x_{ij})+g_2(x_{ij})z_{2ij}+\varepsilon_{ij}, \tag{10.5}$$

其中自变量 z_{2ij} 为虚拟变量, 有 $z_{2ij}=\begin{cases}1, & i=2,\\ 0, & i\neq 2.\end{cases}$

由于 $g_2=f_1-f_2$, 所以检验 $f_1=f_2$ 等同于检验 $g_2=0$, 而 $g(\cdot)$ 是变系数模型 (10.5) 对应的模型系数, 从而将对两条曲线相等性检验问题转化为变系数模型中某一系数函数等于 0 的检验 (即对某一自变量的显著性检验), 可记作

$$H_{03}:\quad g(x_{ij})=0,\quad i=1,2;j=1,2,\cdots,n_i. \tag{10.6}$$

同样检验问题 H_{02} 可转化为变系数模型中某一系数函数是否为常值的检验, 即为

$$H_{04}:\quad g(x_{ij})=a,\quad i=1,2;j=1,2,\cdots,n_i. \tag{10.7}$$

针对检验问题 H_{04}, Cai 等 (2000) 利用 Bootstrap 方法进行检验, Fan 等 (2001) 构造了广义似然比检验统计量, 而 Mei 和 Zhang (2005) 利用方差分析的思想构造了检验统计量. 在此我们基于 Mei 和 Zhang (2005) 的方法来构造检验统计量. 对于变系数模型 (10.5), 将利用基于局部线性光滑的局部加权最小二乘方法进行估计, 用 $\hat{g}_2(x)$ 表示系数 $g_2(x)$ 对应的估计.

针对 H_{03}, H_{04}, 分别构造检验统计量如下

$$T_1=\frac{\dfrac{1}{n_1+n_2}\displaystyle\sum_{i=1}^{2}\sum_{j=1}^{n_i}\hat{g}_2^2(x_{ij})}{\hat{s}^2}, \tag{10.8}$$

$$T_2=\frac{\dfrac{1}{n_1+n_2}\displaystyle\sum_{i=1}^{2}\sum_{j=1}^{n_i}\left\{\hat{g}_2(x_{ij})-\frac{1}{n_1+n_2}\sum_{i=1}^{2}\sum_{j=1}^{n_i}\hat{g}_2(x_{ij})\right\}^2}{\hat{s}^2}, \tag{10.9}$$

其中 $\hat{s}^2$ 是关于模型方差的估计, 可以按照文献 Mei 和 Zhang (2005) 基于模型拟合的残差平方和进行估计, 也可以按照 Dette 和 Neumeyer (2001) 等利用差分方法进行估计.

下面记

$$\boldsymbol{Z}=\begin{pmatrix}\boldsymbol{Z}_1\\ \boldsymbol{Z}_2\end{pmatrix}=\begin{pmatrix}\boldsymbol{z}_{11}^{\mathrm{T}}\\ \vdots\\ \boldsymbol{z}_{1n_1}^{\mathrm{T}}\\ \boldsymbol{z}_{21}^{\mathrm{T}}\\ \vdots\\ \boldsymbol{z}_{2n_2}^{\mathrm{T}}\end{pmatrix}=\begin{pmatrix}1&0\\ \vdots&\vdots\\ 1&0\\ 1&1\\ \vdots&\vdots\\ 1&1\end{pmatrix},\quad \boldsymbol{D}_{x_0}=\begin{pmatrix}\boldsymbol{z}_{11}^{\mathrm{T}}&\boldsymbol{z}_{11}^{\mathrm{T}}(x_{11}-x_0)\\ \vdots&\vdots\\ \boldsymbol{z}_{1n_1}^{\mathrm{T}}&\boldsymbol{z}_{1n_1}^{\mathrm{T}}(x_{1n_1}-x_0)\\ \boldsymbol{z}_{21}^{\mathrm{T}}&\boldsymbol{z}_{21}^{\mathrm{T}}(x_{21}-x_0)\\ \vdots&\vdots\\ \boldsymbol{z}_{2n_2}^{\mathrm{T}}&\boldsymbol{z}_{2n_2}^{\mathrm{T}}(x_{2n_2}-x_0)\end{pmatrix}.$$

对于给定的一点 x_0, 可以基于局部线性方法得到 $(g_1(x_0),g_2(x_0),g_1'(x_0),g_2'(x_0))^{\mathrm{T}}$ 的估计为

$$(\hat{g}_1(x_0),\hat{g}_2(x_0),\hat{g}_1'(x_0),\hat{g}_2'(x_0))^{\mathrm{T}}=\{\boldsymbol{D}_{x_0}^{\mathrm{T}}\boldsymbol{W}_{x_0}\boldsymbol{D}_{x_0}\}^{-1}\boldsymbol{D}_{x_0}^{\mathrm{T}}\boldsymbol{W}_{x_0}\boldsymbol{Y},$$

其中 $\boldsymbol{Y}=(\boldsymbol{Y}_1^{\mathrm{T}},\boldsymbol{Y}_2^{\mathrm{T}})^{\mathrm{T}}$ 和 $\boldsymbol{W}_{x_0}=\mathrm{diag}(\mathrm{K}_h(x_{11}-x_0),\mathrm{K}_h(x_{12}-x_0),\cdots,\mathrm{K}_h(x_{2n_2}-x_0))$, 从而得到 $\boldsymbol{Y}$ 的拟合值为

$$\hat{\boldsymbol{Y}}=\boldsymbol{S}\boldsymbol{Y},$$

其中

$$\boldsymbol{S}=\begin{pmatrix}(\boldsymbol{z}_{11}^{\mathrm{T}},0,0)\{\boldsymbol{D}_{x_{11}}^{\mathrm{T}}\boldsymbol{W}_{x_{11}}\boldsymbol{D}_{x_{11}}\}^{-1}\boldsymbol{D}_{x_{11}}^{\mathrm{T}}\boldsymbol{W}_{x_{11}}\\ (\boldsymbol{z}_{12}^{\mathrm{T}},0,0)\{\boldsymbol{D}_{x_{12}}^{\mathrm{T}}\boldsymbol{W}_{x_{12}}\boldsymbol{D}_{x_{12}}\}^{-1}\boldsymbol{D}_{x_{12}}^{\mathrm{T}}\boldsymbol{W}_{x_{12}}\\ \vdots\\ (\boldsymbol{z}_{2n_2}^{\mathrm{T}},0,0)\{\boldsymbol{D}_{x_{2n_2}}^{\mathrm{T}}\boldsymbol{W}_{x_{2n_2}}\boldsymbol{D}_{x_{2n_2}}\}^{-1}\boldsymbol{D}_{x_{2n_2}}^{\mathrm{T}}\boldsymbol{W}_{x_{2n_2}}\end{pmatrix}.$$

同时可得模型拟合的残差平方和为 $\mathrm{RSS}=(\boldsymbol{Y}-\hat{\boldsymbol{Y}})^{\mathrm{T}}(\boldsymbol{Y}-\hat{\boldsymbol{Y}})=\boldsymbol{Y}^{\mathrm{T}}(\boldsymbol{I}-\boldsymbol{S})^{\mathrm{T}}(\boldsymbol{I}-\boldsymbol{S})\boldsymbol{Y}$, 从而可以以 $\hat{s}^2=\dfrac{\mathrm{RSS}}{n_1+n_2}$ 作为模型误差方差的估计.

上面拟合过程中, 对于窗宽的确定, 采用交叉证实法 (cross-validation). 令

$$\mathrm{CV}(\lambda)=\sum_{j=1}^{2}\sum_{i=1}^{n_j}\left[Y_{ji}-\hat{Y}_{(ji)}(h)\right]^2,$$

其中 $\hat{Y}_{(ji)}(h)$ 为给定 h 值之下, 去掉观测值 x_{ji} 后, 基于其他 n_1+n_2-1 组观测值利用上面所介绍的方法所求得的 Y_{ji} 的估计值. 选择 h, 使得

$$\mathrm{CV}(h_0)=\min\mathrm{CV}(h),$$

则以 h_0 作为参数的估计值. 实际计算中, 可根据经验在 h 的定义域内选择一系列的 h 值分别计算对应的 $\mathrm{CV}(h)$ 的值, 以最小的 $\mathrm{CV}(h)$ 值所对应的 h 值作为 h_0, 或者利用某些优化方法自动选取 h_0.

令

$$M = \begin{pmatrix} e_2^{\mathrm{T}}\{\boldsymbol{D}_{x_{11}}^{\mathrm{T}}\boldsymbol{W}_{x_{11}}\boldsymbol{D}_{x_{11}}\}^{-1}\boldsymbol{D}_{x_{11}}^{\mathrm{T}}\boldsymbol{W}_{x_{11}} \\ e_2^{\mathrm{T}}\{\boldsymbol{D}_{x_{12}}^{\mathrm{T}}\boldsymbol{W}_{x_{12}}\boldsymbol{D}_{x_{12}}\}^{-1}\boldsymbol{D}_{x_{12}}^{\mathrm{T}}\boldsymbol{W}_{x_{12}} \\ \vdots \\ e_2^{\mathrm{T}}\{\boldsymbol{D}_{x_{2n_2}}^{\mathrm{T}}\boldsymbol{W}_{x_{2n_2}}\boldsymbol{D}_{x_{2n_2}}\}^{-1}\boldsymbol{D}_{x_{2n_2}}^{\mathrm{T}}\boldsymbol{W}_{x_{2n_2}} \end{pmatrix},$$

其中 $\boldsymbol{e}_2 = (0,1,0,0)^{\mathrm{T}}$, 从而有 $(\hat{g}_2(x_{11}), \hat{g}_2(x_{12}), \cdots, \hat{g}_2(x_{2n_2}))^{\mathrm{T}} = \boldsymbol{MY}$. 从而可得 T_1 和 T_2 的矩阵形式为

$$T_1 = \frac{\boldsymbol{Y}^{\mathrm{T}}\boldsymbol{M}^{\mathrm{T}}\boldsymbol{MY}}{\boldsymbol{Y}^{\mathrm{T}}(\boldsymbol{I}-\boldsymbol{S})^{\mathrm{T}}(\boldsymbol{I}-\boldsymbol{S})\boldsymbol{Y}},$$

$$T_2 = \frac{\boldsymbol{Y}^{\mathrm{T}}\boldsymbol{M}^{\mathrm{T}}\left(\boldsymbol{I} - \dfrac{1}{n_1+n_2}\boldsymbol{J}\right)\boldsymbol{MY}}{\boldsymbol{Y}^{\mathrm{T}}(\boldsymbol{I}-\boldsymbol{S})^{\mathrm{T}}(\boldsymbol{I}-\boldsymbol{S})\boldsymbol{Y}}.$$

上面的 $\boldsymbol{I}$ 表示 n_1+n_2 维单位矩阵, $\boldsymbol{J}$ 表示元素全为 1 的 n_1+n_2 方阵. 根据基于局部线性估计的变系数模型的理论可知, 在原假设成立的情况下, 对于上面两个统计量近似的有 (在渐近意义上成立)

$$T_1 \approx \frac{\boldsymbol{\varepsilon}^{\mathrm{T}}\boldsymbol{M}^{\mathrm{T}}\boldsymbol{M}\boldsymbol{\varepsilon}}{\boldsymbol{\varepsilon}^{\mathrm{T}}(\boldsymbol{I}-\boldsymbol{S})^{\mathrm{T}}(\boldsymbol{I}-\boldsymbol{S})\boldsymbol{\varepsilon}}, \tag{10.10}$$

和

$$T_2 \approx \frac{\boldsymbol{\varepsilon}^{\mathrm{T}}\boldsymbol{M}^{\mathrm{T}}\left(\boldsymbol{I} - \dfrac{1}{n_1+n_2}\boldsymbol{J}\right)\boldsymbol{M}\boldsymbol{\varepsilon}}{\boldsymbol{\varepsilon}^{\mathrm{T}}(\boldsymbol{I}-\boldsymbol{S})^{\mathrm{T}}(\boldsymbol{I}-\boldsymbol{S})\boldsymbol{\varepsilon}}, \tag{10.11}$$

其中 $\boldsymbol{\varepsilon} = (\boldsymbol{\varepsilon}_1^{\mathrm{T}}, \boldsymbol{\varepsilon}_2^{\mathrm{T}})^{\mathrm{T}}$, 显然对于这两个检验统计量, 可以利用第 1 章中的推论 1.1~推论 1.3 来计算其检验 p 值.

对于上面的检验方法, 有下面两点说明.

(1) 上面将两条曲线设定为一个变系数模型, 从而可以直接利用 Cross-Validation 等方法选择窗宽, 这也是其他文献没有提及的地方.

(2) 我们将证明如下的结论: 如果对于变系数模型利用核方法进行估计, 在窗宽相同的情况下有

$$\hat{g}_2(x) = \hat{f}_1(x) - \hat{f}_2(x),$$

其中 $\hat{f}_1(x), \hat{f}_2(x)$ 是分别只基于第一组数据与第二组数据的估计值. 所以检验统计量 T_1 为

$$T_1 = \frac{\dfrac{1}{n_1+n_2}\displaystyle\sum_{i=1}^{2}\sum_{j=1}^{n_i}(\hat{f}_1(x_{ij}) - \hat{f}_2(x_{ij}))^2}{\hat{s}^2}.$$

如果 $n_1 = n_2 = n, x_{1j} = x_{2j}, j = 1, 2, \cdots, n$, 即两条曲线的自变量相同, 有

$$T_1 = \frac{\dfrac{1}{n}\sum_{i=1}^{n}\{\hat{f}_1(x_i) - \hat{f}_2(x_i)\}^2}{\hat{s}^2}.$$

显然此时 T_1 就是 King 等 (1991) 所构造的检验统计量. 所以我们构造的检验方法是 King 等 (1991) 所提检验方法在曲线的自变量不同时的推广.

上面结论的证明　首先有

$$\boldsymbol{Z} = \begin{pmatrix} \mathbf{1}_{n_1} & \mathbf{0}_{n_1} \\ \mathbf{1}_{n_2} & \mathbf{1}_{n_2} \end{pmatrix},$$

令 $\boldsymbol{W}_i = \mathrm{diag}\left(K_h(x_{i1} - x_0), K_h(x_{i2} - x_0), \cdots, K_h(x_{in_1} - x_0)\right), i = 1, 2$, 所以可得

$$\begin{aligned}(\hat{g}_1(x_0), \hat{g}_2(x_0))^{\mathrm{T}} &= (\boldsymbol{Z}^{\mathrm{T}}\boldsymbol{W}\boldsymbol{Z})^{-1}\boldsymbol{Z}^{\mathrm{T}}\boldsymbol{W}\boldsymbol{Y} \\ &= \begin{pmatrix} (\mathbf{1}_{n_1}^{\mathrm{T}}\boldsymbol{W}_1\mathbf{1}_{n_1})^{-1}\mathbf{1}_{n_1}^{\mathrm{T}}\boldsymbol{W}_1\boldsymbol{Y}_1 \\ (\mathbf{1}_{n_2}^{\mathrm{T}}\boldsymbol{W}_2\mathbf{1}_{n_2})^{-1}\mathbf{1}_{n_2}^{\mathrm{T}}\boldsymbol{W}_2\boldsymbol{Y}_2 - (\mathbf{1}_{n_1}^{\mathrm{T}}\boldsymbol{W}_1\mathbf{1}_{n_1})^{-1}\mathbf{1}_{n_1}^{\mathrm{T}}\boldsymbol{W}_1\boldsymbol{Y}_1 \end{pmatrix} \\ &= \begin{pmatrix} \hat{f}_1(x_0) \\ \hat{f}_2(x_0) - \hat{f}_1(x_0) \end{pmatrix},\end{aligned}$$

从而

$$T_1 = \sum_{i=1}^{2}\sum_{j=1}^{n_i}\hat{g}_2^2(x_{ij}) = \sum_{i=1}^{2}\sum_{j=1}^{n_i}(\hat{f}_2(x_{ij}) - \hat{f}_1(x_{ij}))^2.$$

显然 T_1 为 King 等 (1991) 在自变量不同情形下的推广.

数值模拟　下面将通过数值模拟来考察本节所提虚拟变量检验方法的有效性. 考虑如下的模型

$$\begin{cases} y_{1i} = f_1(x_{1i}) + \varepsilon_{1i}, & \varepsilon_{1i} \sim N(0, \sigma_1^2), \quad i = 1, 2, \cdots, n_1, \\ y_{2j} = f_2(x_{2j}) + \varepsilon_{2j}, & \varepsilon_{2j} \sim N(0, \sigma_2^2), \quad j = 1, 2, \cdots, n_2. \end{cases}$$

为了全面考察检验方法的有效性, 对回归函数取下面三种情形.

$$\begin{aligned} &M_1: \quad f_1 = x; \quad f_2 = \beta x; \quad \beta = 0.5, 1, 1.5, \\ &M_2: \quad f_1 = 0; \quad f_2 = \beta\cos(2\pi x); \quad \beta = 0, 0.5, 1, \\ &M_3: \quad f_1 = x; \quad f_2 = \beta x + 1; \quad \beta = 0, 1, 2, \end{aligned}$$

其中针对 M_1 和 M_2 来检验两条曲线是否相同, 针对 M_3 检验两条曲线是否平行. 所采用的检验统计量为式 (10.8) 和式 (10.9) 形式的虚拟变量检验统计量. 对样本量 n_1, n_2, 误差方差 σ_1, σ_2 取下面四种情形

$$(\mathrm{A.1}): n_1 = n_2 = 30, \sigma_1 = \sigma_2 = 0.2; \quad (\mathrm{A.2}): n_1 = n_2 = 30, 2\sigma_1 = \sigma_2 = 0.2$$

$$(\text{A.3}): n_1 = 1.5n_2 = 30, \sigma_1 = \sigma_2 = 0.2; \quad (\text{A.4}): n_1 = 1.5n_2 = 30, 2\sigma_1 = \sigma_2 = 0.2$$

模拟中 x 是服从 $U(0,1)$ 分布的随机数. 对于上面的每种情况, 取显著水平 $\alpha = 0.05$, 以 500 次重复中检验 p 值小于 α (即拒绝 H_0) 的频率模拟检验功效, 结果见表 10.1. 从模拟结果可以看出:

(1) 虚拟变量检验方法对于检验两条曲线是否相等以及平行具有很高的灵敏度. 当原假设成立时, 拒绝 H_0 的频率一般接近显著水平 $\alpha = 0.05$; 而当原假设不成时, 拒绝 H_0 的频率 (检验功效) 一般接近 1.

(2) 样本量的变化为对检验精度影响不大, 而误差方差的变化对检验精度有显著的影响.

10.3 光滑残差检验法

残差分析是统计诊断中最为常用与直接的方法, 随着非参数回归技术的发展, 近年来, 基于残差分析的思想借助于光滑方法来构造检验统计量受到了人们的重视, 其中包括 Hart (1997) 研究的光滑残差法 (smoothing residuals) 和 Azzalini 和 Bowman (1993), Dette (2000) 研究的伪似然比方法 (pseudo-likelihood ratio). 下面利用该思想来研究非参数协方差分析模型.

本节考虑非参数协方差分析模型 (10.1) 的检验问题 (10.2). 即研究对 $Y_{ij} = f_i(x_{ij}) + \varepsilon_{ij}$, $i = 1, 2, \cdots, p$, $j = 1, 2, \cdots, n_i$ 的检验问题

$$H_0: \quad f_1 = f_2 = \cdots = f_p, \quad \text{VS} \quad H_1: \quad f_i \neq f_j \quad \text{对某些 } i,\ j \in \{1, \cdots, p\}.$$

如果原假设 H_0 成立, 即所有的 p 条回归曲线都相等, 上面的模型变为

$$Y_{ij} = f(x_{ij}) + \varepsilon_{ij}.$$

下面基于局部线性方法利用所有的数据 (合并在一起的数据) 来估计未知回归函数 $f(\cdot)$, 给定的一点 x_{ij}, 令

$$\boldsymbol{Y} = \begin{pmatrix} y_{11} \\ y_{12} \\ \vdots \\ y_{pn_p} \end{pmatrix}, \quad \boldsymbol{D}_{x_{ij}} = \begin{pmatrix} 1 & x_{11} - x_{ij} \\ 1 & x_{12} - x_{ij} \\ \vdots & \vdots \\ 1 & x_{pn_p} - x_{ij} \end{pmatrix},$$

以及 $\boldsymbol{W}_{x_{ij}} = \text{diag}(K_h(x_{11} - x_{ij}), K_h(x_{12} - x_{ij}), \cdots, K_h(x_{pn_p} - x_{ij}))$, 其中 $K_h(\cdot) = K(\cdot/h)/h$, K 是核函数, h 是窗宽. 从而有 $f(x_{ij})$ 的 (合成) 估计为

$$\hat{f}(x_{ij}) = (1 \ \ 0)\{\boldsymbol{D}_{x_{ij}}^{\text{T}} \boldsymbol{W}_{x_{ij}} \boldsymbol{D}_{x_{ij}}\}^{-1} \boldsymbol{D}_{x_{ij}}^{\text{T}} \boldsymbol{W}_{x_{ij}} \boldsymbol{Y}.$$

相应的拟合残差可记为

$$\begin{aligned}\hat{e}_{ij} &= y_{ij} - \hat{f}(x_{ij}) = y_{ij} - f_i(x_{ij}) + f_i(x_{ij}) - f(x_{ij}) + f(x_{ij}) - \hat{f}(x_{ij}) \\ &= \varepsilon_{ij} + [f_i(x_{ij}) - f(x_{ij})] + [f(x_{ij}) - \hat{f}(x_{ij})]. \end{aligned} \tag{10.12}$$

从式 (10.12) 可以看到, 如果 H_0 成立, 即 $f_i = f$, 又由局部线性估计的性质知在一定条件下 $\hat{f}(x_{ij}) \to f(x_{ij})$, 从而有 $\hat{e}_{ij}$ 趋于 ε_{ij}. 但是在备择假设 H_1 下, 即 $f_i \neq f$, $\hat{e}_{ij}$ 趋于 $\varepsilon_{ij} + [f_i(x_{ij}) - f(x_{ij})]$.

基于上面的分析, 可以认为 $\hat{e}_{ij}$ 为 x_{ij} 的光滑函数, 记做

$$\hat{e}_{ij} = m(x_{ij}) + \eta_{ij}. \tag{10.13}$$

类似于上面 $f(x_{ij})$ 的估计, 同样利用局部线性方法来估计 $m(x_{ij})$, 从而有

$$\hat{m}(x_{ij}) = (1\ \ 0)\{\boldsymbol{D}_{x_{ij}}^{\mathrm{T}} \boldsymbol{W}_{x_{ij}}^{*} \boldsymbol{D}_{x_{ij}}\}^{-1} \boldsymbol{D}_{x_{ij}}^{\mathrm{T}} \boldsymbol{W}_{x_{ij}}^{*} \hat{\boldsymbol{e}},$$

其中 $\hat{\boldsymbol{e}} = (\hat{e}_{11}, \hat{e}_{12}, \cdots, \hat{e}_{pn_p})^{\mathrm{T}}$, $W_{x_{ij}}^{*}$ 与 $W_{x_{ij}}$ 形式一样, 只是窗宽由 h 变为 h^{*}.

类似于 Hart (1997) 研究的光滑残差法, 记 $\sum_{i=1}^{p} n_i = N$, 构造如下的检验统计量

$$T = \frac{\dfrac{1}{N}\sum_{i=1}^{p}\sum_{j=1}^{n_i} \hat{m}^2(x_{ij})}{\hat{s}^2}, \tag{10.14}$$

其中 $\hat{s}^2$ 是合并数据在备择假设下方差的估计. 基于上面关于 $\hat{e}_{ij}$ 的分析可知, 如果 T 值较大, 倾向于拒绝原假设. 令 T 的观测值为 t, 则 T 的检验 p 值为

$$p_0 = P_{H_0}\ (T > t), \tag{10.15}$$

即在 H_0 之下, 事件 $\{T > t\}$ 发生的概率. 对于给定的显著水平 α, 若 $p_0 > \alpha$, 则接受 H_0; 若 $p_0 \leqslant \alpha$, 则拒绝 $\mathrm{H_0}$.

令

$$\boldsymbol{S} = \begin{pmatrix} (1\ \ 0)\{\boldsymbol{D}_{x_{11}}^{\mathrm{T}} \boldsymbol{W}_{x_{11}} \boldsymbol{D}_{x_{11}}\}^{-1} \boldsymbol{D}_{x_{11}}^{\mathrm{T}} \boldsymbol{W}_{x_{11}} \\ (1\ \ 0)\{\boldsymbol{D}_{x_{12}}^{\mathrm{T}} \boldsymbol{W}_{x_{12}} \boldsymbol{D}_{x_{12}}\}^{-1} \boldsymbol{D}_{x_{12}}^{\mathrm{T}} \boldsymbol{W}_{x_{12}} \\ \vdots \\ (1\ \ 0)\{\boldsymbol{D}_{x_{pn_p}}^{\mathrm{T}} \boldsymbol{W}_{x_{pn_p}} \boldsymbol{D}_{x_{pn_p}}\}^{-1} \boldsymbol{D}_{x_{pn_p}}^{\mathrm{T}} \boldsymbol{W}_{x_{pn_p}} \end{pmatrix},$$

对应的定义 $\boldsymbol{S}^{*}$, 即将 $\boldsymbol{S}$ 中的 $W_{x_{ij}}$ 该作 $W_{x_{ij}}^{*}$, 则有

$$\sum_{i=1}^{p}\sum_{j=1}^{n_i} \hat{m}^2(x_{ij}) = \boldsymbol{Y}^{\mathrm{T}}(\boldsymbol{I} - \boldsymbol{S})^{\mathrm{T}} \boldsymbol{S}^{*\mathrm{T}} \boldsymbol{S}^{*} (\boldsymbol{I} - \boldsymbol{S})\boldsymbol{Y}. \tag{10.16}$$

至于 $\hat{s}^2$, 可以像 King, Hart 和 Wehrly (1991) 与 Young 和 Bowman (1995) 那样利用差分方法进行估计, 或者基于拟合的残差平方和进行估计. 这两种估计方法得到的 $\hat{s}^2$ 都可是如下的形式

$$\hat{s}^2 = \boldsymbol{Y}^{\mathrm{T}}(\boldsymbol{I}-\boldsymbol{L})^{\mathrm{T}}(\boldsymbol{I}-\boldsymbol{L})\boldsymbol{Y}. \tag{10.17}$$

由式 (10.16) 和式 (10.17), 将 T 写成下面的矩阵形式

$$T = \frac{\boldsymbol{Y}^{\mathrm{T}}(\boldsymbol{I}-\boldsymbol{S})^{\mathrm{T}}\boldsymbol{S}^{*\mathrm{T}}\boldsymbol{S}^*(\boldsymbol{I}-\boldsymbol{S})\boldsymbol{Y}}{\boldsymbol{Y}^{\mathrm{T}}(\boldsymbol{I}-\boldsymbol{L})^{\mathrm{T}}(\boldsymbol{I}-\boldsymbol{L})\boldsymbol{Y}}. \tag{10.18}$$

正如 King, Hart 和 Wehrly (1991) 与 Young 和 Bowman (1995) 的分析, 去掉趋于零的部分, 上面的统计量近似的有

$$T = \frac{\boldsymbol{\varepsilon}^{\mathrm{T}}(\boldsymbol{I}-\boldsymbol{S})^{\mathrm{T}}\boldsymbol{S}^{*\mathrm{T}}\boldsymbol{S}^*(\boldsymbol{I}-\boldsymbol{S})\boldsymbol{\varepsilon}}{\boldsymbol{\varepsilon}^{\mathrm{T}}(\boldsymbol{I}-\boldsymbol{L})^{\mathrm{T}}(\boldsymbol{I}-\boldsymbol{L})\boldsymbol{\varepsilon}}.$$

对于这类检验统计量, 显然可以利用第 1 章中的推论 1.1~ 推论 1.3 来计算其检验 p 值.

值得注意的是, 基于式 (10.12) 中 $\hat{e}_{ij}$ 的分析, 也可以构造 Azzalini 和 Bowman (1991) 类型的拟似然比检验, 形式如下

$$T = \frac{\sum_{i=1}^{p}\sum_{j=1}^{n_i}\hat{e}_{ij}^2 - \sum_{i=1}^{p}\sum_{j=1}^{n_i}[\hat{e}_{ij}-\hat{m}(x_{ij})]^2}{\sum_{i=1}^{p}\sum_{j=1}^{n_i}[\hat{e}_{ij}-\hat{m}(x_{ij})]^2}. \tag{10.19}$$

显然, 如果原假设成立, 基于上面的分析有 $\hat{e}_{ij}$ 趋于 ε_{ij}, 那么 $\hat{e}_{ij}$ 与 $\hat{e}_{ij}-\hat{m}(x_{ij})$ 没有显著差别, 否则二者差别显著. 所以 T 越大, 越倾向于拒绝原假设.

10.4 个体差异曲线检验法

类似于传统方差分析模型的分析, 下面对于一类特殊的非参数协方差分析模型重新设定, 求出总体均值曲线以及个体差异曲线的估计, 并且基于差异曲线的估计构造了检验统计量. 本节内容主要来自于文献魏传华和吴喜之 (2006).

对于如下的非参数协方差分析模型

$$Y_{ij} = f_i(x_j) + \varepsilon_{ij}, \quad i=1,2,\cdots,p; \quad j=1,2,\cdots,n. \tag{10.20}$$

模型 (10.20) 可写为下面的形式

$$\begin{cases} Y_{ij} = g(x_j) + g_i(x_j) + \varepsilon_{ij}, \\ \sum_{i=1}^{p} g_i(x) = 0, \quad \forall x, \end{cases} \tag{10.21}$$

其中定义 $g(\cdot)$ 为总体均值曲线, $g_i(\cdot)$ 为个体差异曲线. 那么检验问题 (10.2) 可等价地写为

$$H_0: \quad g_1(x) = g_2(x) = \cdots = g_p(x) = 0, \quad \forall x. \tag{10.22}$$

如果记 $g_i(x_j)$ 的估计为 $\hat{g}_i(x_j)$, 那么可以构造如下的检验统计量

$$T = \frac{\sum_{i=1}^{p}\sum_{j=1}^{n} \hat{g}_i^2(x_j)}{\hat{s}^2}. \tag{10.23}$$

值得注意的是, 如果各条曲线对于 $\forall x$, 有 $f_i(x) = \alpha_i$, 显然上面的模型就是经典的方差分析模型.

模型 (10.21) 显然是一个含约束条件的变系数模型, 关于这类模型的估计可按含约束条件的线性模型的估计方法进行. 下面基于核方法进行估计.

首先模型 (10.21) 可记为如下形式的变系数模型

$$y_{ij} = g(x_j) + g_1(x_j)z_{1ij} + g_2(x_j)z_{2ij} + \cdots + g_p(x_j)z_{pij} + \varepsilon_{ij}, \tag{10.24}$$

其中自变量 z_{kij} 为虚拟变量, 有 $z_{kij} = \begin{cases} 1, & i = k, \\ 0, & i \neq k, \end{cases}$ $k = 1, 2, \cdots, p$. 对于指定的点 x_0, $g(x_0), g_i(x_0)$ 可通过使得

$$\sum_{i=1}^{p}\sum_{j=1}^{n} \{y_{ij} - g(x_0) - g_1(x_0)z_{1ij} - \cdots - g_p(x_0)z_{pij}\}^2 K_h(x_j - x_0) \tag{10.25}$$

达到最小予以估计. 其中 $K_h(\cdot) = K(\cdot/h)/h$, K 是核函数, h 是窗宽.

令

$$\boldsymbol{A}_1 = (\mathbf{1}_n^{\mathrm{T}}, \mathbf{1}_n^{\mathrm{T}}, \cdots, \mathbf{1}_n^{\mathrm{T}})^{\mathrm{T}}, \quad \boldsymbol{A}_2 = \mathrm{diag}\{\mathbf{1}_n, \mathbf{1}_n, \cdots, \mathbf{1}_n\}_{np\times p}, \ \boldsymbol{Z} = (\boldsymbol{A}_1, \boldsymbol{A}_2),$$

$$\boldsymbol{G}(x_0) = (g(x_0), g_1(x_0), \cdots, g_p(x_0))^{\mathrm{T}}, \quad \boldsymbol{g}(x_0) = (g_1(x_0), \cdots, g_p(x_0))^{\mathrm{T}},$$

$$\boldsymbol{K}(x_0) = \mathrm{diag}(K_h(x_1 - x_0), K_h(x_2 - x_0), \cdots, K_h(x_n - x_0)),$$

$$\boldsymbol{W} = \mathrm{diag}(\boldsymbol{K}(x_0), \boldsymbol{K}(x_0), \cdots, \boldsymbol{K}(x_0)).$$

基于式 (10.25) 的计算可得如下的估计方程

$$\boldsymbol{Z}^{\mathrm{T}}\boldsymbol{W}\boldsymbol{Z}\boldsymbol{G}(x_0)=\boldsymbol{Z}^{\mathrm{T}}\boldsymbol{W}\boldsymbol{Y}, \tag{10.26}$$

即为

$$\begin{pmatrix} \boldsymbol{A}_1^{\mathrm{T}}\boldsymbol{W}\boldsymbol{A}_1 & \boldsymbol{A}_1^{\mathrm{T}}\boldsymbol{W}\boldsymbol{A}_2 \\ \boldsymbol{A}_2^{\mathrm{T}}\boldsymbol{W}\boldsymbol{A}_1 & \boldsymbol{A}_2^{\mathrm{T}}\boldsymbol{W}\boldsymbol{A}_2 \end{pmatrix}\begin{pmatrix} g_0(x_0) \\ \boldsymbol{g}(x_0) \end{pmatrix}=\begin{pmatrix} \boldsymbol{A}_1^{\mathrm{T}}\boldsymbol{W}\boldsymbol{Y} \\ \boldsymbol{A}_1^{\mathrm{T}}\boldsymbol{W}\boldsymbol{Y} \end{pmatrix},$$

等价地有

$$\begin{cases} \boldsymbol{A}_1^{\mathrm{T}}\boldsymbol{W}\boldsymbol{A}_1 g_0(x_0)+\boldsymbol{A}_1^{\mathrm{T}}\boldsymbol{W}\boldsymbol{A}_2\boldsymbol{g}(x_0)=\boldsymbol{A}_1^{\mathrm{T}}\boldsymbol{W}\boldsymbol{Y}, \\ \boldsymbol{A}_2^{\mathrm{T}}\boldsymbol{W}\boldsymbol{A}_1 g_0(x_0)+\boldsymbol{A}_2^{\mathrm{T}}\boldsymbol{W}\boldsymbol{A}_2\boldsymbol{g}(x_0)=\boldsymbol{A}_1^{\mathrm{T}}\boldsymbol{W}\boldsymbol{Y}. \end{cases} \tag{10.27}$$

式 (10.27) 中的第一式可化为

$$p\sum_{j=1}^{n}K_h(x_j-x_0)g_0(x_0)+\sum_{i=1}^{p}\sum_{j=1}^{n}K_h(x_j-x_0)g_i(x_0)=\sum_{i=1}^{p}\sum_{j=1}^{n}K_h(x_j-x_0)y_{ij}, \tag{10.28}$$

第二式为

$$\sum_{j=1}^{n}K_h(x_j-x_0)g_0(x_0)+\sum_{j=1}^{n}K_h(x_j-x_0)g_i(x_0)=\sum_{j=1}^{n}K_h(x_j-x_0)y_{ij},\quad i=1,2,\cdots,p, \tag{10.29}$$

再加上条件

$$\sum_{i=1}^{p}g_i(x_0)=0.$$

可求得

$$\hat{g}_0(x_0)=\frac{\sum_{i=1}^{2}\sum_{j=1}^{n}K_h(x_j-x_0)y_{ij}}{p\sum_{j=1}^{n}K_h(x_j-x_0)}=\hat{f}(x_0). \tag{10.30}$$

$$\hat{g}_i(x_0)=\frac{\sum_{j=1}^{n}K_h(x_j-x_0)y_{ij}}{\sum_{j=1}^{n}K_h(x_j-x_0)}-\hat{g}_0(x_0)=\hat{f}_i(x_0)-\hat{f}(x_0), \tag{10.31}$$

其中 $\hat{f}(x_0)$ 是基于合成数据 $\{y_{ij},x_j\}(i=1,\cdots,p;j=1,\cdots,p)$ 利用核方法估计所得, $\hat{f}_i(x_0)$ 是基于各自的数据 $\{y_{ij},x_j\}(j=1,\cdots,p)$ 利用核方法估计所得. 这个结论也与经典的方差分析模型相一致. 对于所构造的检验统计量有

$$\sum_{i=1}^{p}\sum_{j=1}^{n}\hat{g}_i^2(x_j)=\sum_{i=1}^{p}\sum_{j=1}^{n}\{\hat{f}_i(x_j)-\hat{f}(x_j)\}^2.$$

注意 Young 和 Bowman(1995) 对于一般的协方差分析模型 (10.1) 的检验问题 (10.2) 构造了如下形式的检验统计量

$$T_{\mathrm{YB}} = \frac{\sum_{i=1}^{p}\sum_{j=1}^{n_i}(\hat{f}_i(x_{ij}) - \hat{f}(x_{ij}))^2}{\hat{s}^2}.$$

因此我们构造的检验统计量是 Young 和 Bowman (1995) 所提方法在自变量相同情形下的推广.

同 Young 和 Bowman (1995) 一样, 可近似地将检验统计量化为误差变量 ε 变量二次型之比的形式, 从而利用第 1 章中的推论 1.1~ 推论 1.3 来计算其检验 p 值.

值得注意的是, 如果模型 (10.20) 中 $p=2$, 即研究两条曲线在自变量相同情况下的的比较. 上面的方法与 Young 和 Bowman (1995) 主要是考察

$$\boldsymbol{A}_1 = \sum_{i=1}^{2}\sum_{j=1}^{n}\{\hat{f}_i(x_j) - \hat{f}(x_j)\}^2.$$

而第二节的虚拟变量法与 King, Hart 和 Wehrly (1991) 考察的是

$$\boldsymbol{A}_2 = \sum_{i=1}^{n}\{\hat{f}_1(x_i) - \hat{f}_2(x_i)\}^2.$$

对于 $\boldsymbol{A}_1$, $\boldsymbol{A}_2$ (窗宽相等情况下) 二者有如下的关系

$$2\boldsymbol{A}_1 = \boldsymbol{A}_2. \tag{10.32}$$

下面我们给出简要证明. 原假设成立时, 对于给定的一点 x_0, 基于核方法有

$$\hat{f}(x_0) = \frac{\sum_{i=1}^{2}\sum_{j=1}^{n}K_h(x_j - x_0)y_{ij}}{\sum_{i=1}^{2}\sum_{j=1}^{n}K_h(x_j - x_0)} = \frac{\sum_{i=1}^{2}\sum_{j=1}^{n}K_h(x_j - x_0)y_{ij}}{2\sum_{j=1}^{n}K_h(x_j - x_0)}.$$

而备择假设下有

$$\hat{f}_1(x_0) = \frac{\sum_{j=1}^{n}K_h(x_j - x_0)y_{1j}}{\sum_{j=1}^{n}K_h(x_j - x_0)},$$

与

$$\hat{f}_2(x_0)=\frac{\sum_{j=1}^{n}K_h(x_j-x_0)y_{2j}}{\sum_{j=1}^{n}K_h(x_j-x_0)}.$$

显然有

$$\hat{f}_1(x_0)+\hat{f}_2(x_0)=2\hat{f}(x_0),$$

即

$$\hat{f}_1(x_0)-\hat{f}(x_0)=\hat{f}(x_0)-\hat{f}_2(x_0).$$

所以有

$$\begin{aligned}\boldsymbol{A}_1&=\sum_{i=1}^{2}\sum_{j=1}^{n}(\hat{f}_i(x_j)-\hat{f}(x_j))^2\\&=\sum_{j=1}^{n}(\hat{f}_1(x_j)-\hat{f}(x_j))^2+\sum_{j=1}^{n}(\hat{f}_2(x_j)-\hat{f}(x_j))^2\\&=2\sum_{j=1}^{n}(\hat{f}(x_j)-\hat{f}_2(x_j))^2,\end{aligned}$$

以及

$$\begin{aligned}\boldsymbol{A}_2&=\sum_{j=1}^{n}(\hat{f}_1(x_j)-\hat{f}_2(x_j))^2\\&=\sum_{j=1}^{n}(\hat{f}_1(x_j)+\hat{f}_2(x_j)-2\hat{f}_2(x_j))^2\\&=4\sum_{j=1}^{n}(\hat{f}(x_j)-\hat{f}_2(x_j))^2.\end{aligned}$$

因此有

$$2\boldsymbol{A}_1=\boldsymbol{A}_2.$$

前面基于变系数模型的思想提出了虚拟变量法和个体差异曲线法, 但是其中的虚拟变量法主要是侧重于检验两条曲线, 而个体差异曲线法主要针对自变量相同的多条曲线的比较. 对于一般情况下多条非参数回归曲线的比较, 我们仍然可以在变系数模型的框架下进行分析. 注意到备择假设下可将非参数协方差分析模型设定为一般的变系数模型, 而原假设成立时, 模型是一个一般的非参数回归模型. 对于这类非参数模型对非参数模型的检验. Fan, Zhang 和 Zhang (2001) 提出了广义似然比检验统计量, 并研究了和变系数模型有关的几类检验. 并且都证明了该类检验统计量在原假设成立时为 χ^2 分布, 即存在 Wilks 现象, 得到了较为深刻的结果. 虽然上面的检验问题也能归到 Fan, Zhang 和 Zhang (2001) 研究的问题中, 但是由于非参数协方差分析模型的特殊性, 下面专门讨论模型 (10.1) 的广义似然比检验.

首先在备择假设下, 即对 p 条回归曲线没有任何约束, 可将模型 (10.1) 记为

$$\begin{pmatrix} \boldsymbol{Y}_1 \\ \boldsymbol{Y}_2 \\ \vdots \\ \boldsymbol{Y}_p \end{pmatrix} = \begin{pmatrix} f_1(\boldsymbol{x}_1) \\ \mathbf{0}_{n_2} \\ \vdots \\ \mathbf{0}_{n_p} \end{pmatrix} + \begin{pmatrix} \mathbf{0}_{n_1} \\ f_2(\boldsymbol{x}_2) \\ \vdots \\ \mathbf{0}_{n_p} \end{pmatrix} + \cdots + \begin{pmatrix} \mathbf{0}_{n_1} \\ \mathbf{0}_{n_2} \\ \vdots \\ f_p(\boldsymbol{x}_p) \end{pmatrix} + \begin{pmatrix} \boldsymbol{\varepsilon}_1 \\ \boldsymbol{\varepsilon}_2 \\ \vdots \\ \boldsymbol{\varepsilon}_p \end{pmatrix},$$

其中

$$\boldsymbol{Y}_i = \begin{pmatrix} y_{i1} \\ y_{i2} \\ \vdots \\ y_{in_i} \end{pmatrix}, \quad f_i(\boldsymbol{x}_i) = \begin{pmatrix} f_i(x_{i1}) \\ f_i(x_{i2}) \\ \vdots \\ f_i(x_{in_i}) \end{pmatrix}, \quad \boldsymbol{\varepsilon}_i = \begin{pmatrix} \varepsilon_{i1} \\ \varepsilon_{i2} \\ \vdots \\ \varepsilon_{in_i} \end{pmatrix}, \quad i = 1, 2, \cdots, p.$$

显然上面的模型为一变系数模型, 可记作

$$y_{ij} = f_1(x_{ij})z_{1ij} + f_2(x_{ij})z_{2ij} + \cdots + f_p(x_{ij})z_{pij} + \varepsilon_{ij}, \tag{10.33}$$

其中自变量 z_{kij} 为虚拟变量, 有 $z_{kij} = \begin{cases} 1, & i = k, \\ 0, & i \neq k, \end{cases}$ $k = 1, 2, \cdots, p$. 从而我们将多条非参数回归曲线的检验问题 (10.2) 转变为变系数模型 (10.33) 中关于模型系数函数的检验问题.

对于变系数模型 (10.33), 采用局部线性方法进行估计, 记 $f_i(x)$ 估计的结果为 $\hat{f}_i(x)$. 可以证明该估计与各条曲线各自单独估计得到的结果形式一样, 但是基于模型 (10.33) 进行估计可以直接采用窗宽选择方法. 如果原假设成立, 即有 $f_i = f$, 利用所有 (合成) 数据估计 $f(x)$, 记为 $\hat{f}(x)$. 下面给出广义似然比检验统计量的构造过程.

首先假定 $\boldsymbol{\varepsilon} \sim N(\mathbf{0}, \sigma^2 \boldsymbol{I}_n)$, 那么模型 (10.2) 的对数似然函数为

$$L = -\frac{N}{2}\log 2\pi\sigma^2 - \frac{1}{2\sigma^2}\sum_{i=1}^{p}\sum_{j=1}^{n_i}\{y_{ij} - f_1(x_{ij})z_{1ij} - f_2(x_{ij})z_{2ij} - \cdots - f_p(x_{ij})z_{pij}\}^2, \tag{10.34}$$

则可求得 x 处对应的 $f_i(x)$ 的局部似然估计以及 σ^2 的估计, 其中 $f_i(x)$ 的局部似然估计等于 $\hat{f}_i(x)$, 最后可得如下的似然函数

$$L(H_1) = -\frac{n}{2} - \frac{n}{2}\log 2\pi/n - \frac{n}{2}\log \mathrm{RSS}_1, \tag{10.35}$$

其中

$$\begin{aligned}\mathrm{RSS}_1 &= \sum_{i=1}^{p}\sum_{j=1}^{n_i}\left\{y_{ij}-\hat{f}_1(x_{ij})z_{1ij}-\hat{f}_2(x_{ij})z_{2ij}-\cdots-\hat{f}_p(x_{ij})z_{pij}\right\}^2\\ &= \sum_{i=1}^{p}\sum_{j=1}^{n_i}\left\{y_{ij}-\hat{f}_i(x_{ij})\right\}^2.\end{aligned}$$

同样原假设成立时, 最后可得似然函数为

$$L(H_0) = -\frac{n}{2}-\frac{n}{2}\log 2\pi/n-\frac{n}{2}\log\mathrm{RSS}_0, \tag{10.36}$$

其中

$$\mathrm{RSS}_0 = \sum_{i=1}^{p}\sum_{j=1}^{n_i}\left\{y_{ij}-\hat{f}(x_{ij})\right\}^2.$$

基于式 (10.35) 与式 (10.36), 构造广义似然比检验检验量

$$\mathrm{T_{GLR}} = L(H_1)-L(H_0) = \frac{n}{2}\log\frac{\mathrm{RSS}_0}{\mathrm{RSS}_1} \approx \frac{n}{2}\frac{\mathrm{RSS}_0-\mathrm{RSS}_1}{\mathrm{RSS}_1}. \tag{10.37}$$

在误差为正态分布的假定下, 可以利用推论 1.1~ 推论 1.3 来计算其检验 p 值.

关于该检验统计量的渐近性质, 可参考 Fan, Zhang 和 Zhang (2001) 以及 Dette 和 Neumeyer (2001) 的结果.

此外, 本节主要讨论的是非参数曲线的比较. 众所周知, 当自变量的维数较大时, 用简单的非参数回归方法拟合实际数据效果不好. 半参数建模方法作为适当的折中可以解决这一问题. 魏传华和吴喜之 (2008a) 研究了两类部分线性模型的比较问题.

参 考 文 献

柴根象, 洪圣岩. 1995. 半参数回归模型. 合肥：安徽教育出版社.

梅长林. 2000. 泛函系数回归模型的研究及其应用. 西安交通大学博士学位论文.

王启华, 史宁中, 耿直. 2010. 现代统计研究基础. 北京：科学出版社.

魏传华, 李静, 吴喜之. 2009. 部分线性模型基于参数信息的统计推断. 数学的实践与认识, 39(19): 162-168.

魏传华, 梅长林. 2005. 半参数空间变系数回归模型的两步估计方法及数值模拟. 统计与信息论坛, (1): 17-19.

魏传华, 梅长林. 2006. 半参数空间变系数回归模型的 Back-Fitting 估计. 数学的实践与认识, (3): 177-184.

魏传华, 吴喜之. 2006. 非参数协方差分析基于变系数模型的统计推断. 数学的实践与认识, (12): 208-215.

魏传华, 吴喜之. 2007. Testing Linearity for Nonparametric Component of Partially Linear Models. 应用数学, (1): 183-190.

魏传华, 吴喜之. 2008a. 两个部分线性模型的比较. 统计研究, (1): 82-85.

魏传华, 吴喜之. 2008b. 部分线性变系数模型 Backfitting 估计的渐近性质. 高校应用数学学报, (3): 227-234.

魏传华, 吴喜之. 2008c. 部分线性变系数模型的 profile Lagrange 乘子检验. 系统科学与数学, (4): 416-424.

魏传华, 吴喜之. 2008d. 部分线性模型非参数部分的线性关系检验. 工程数学学报, 25(5): 857-866.

魏传华. 2010. 因变量缺失下部分线性变系数变量含误差模型的估计. 数学物理学报, 30A(4): 1042-1054.

Ahmad I, Leelahanon S, Li Q. 2005. Efficient estimation of Semiparametric partially linear varying coefficient model. The Annals of Statistics, 33: 258-283.

AlcalGa J T, ChristGobal J A, GonzGalez Manteiga W. 1999. Goodness-of-t test for linear models based on local polynomials. Statistics and Probability Letters, 42: 39-46.

Ali M A, Abu-Salih M S. 1988. On estimation of missing observations in linear regression models. Sankhya. Indian Journal of Statist, 50(B): 404-411.

Anselin L. 1988. Spatial Econometrics: Methods and Models. Dordrecht: Kluwer Academic.

Azzalini A, Bowman A. 1993. On the use of nonparametric regression for checking linear relationships. Journal of the Royal Statistical Sociation Series B, 55: 549-559.

Bates D M, Watts D G. 1988. Nonlinear Regression Analysis and Its Applications. New York: Wiley.

Belsley D A, Kuh E, Welsch R E. 1980. Regression Diagnostics: Identifying Influential Data and Sources of Collinearity. New York: Wiley.

Beriman L, Meisel W, Purcell E. 1977. Variable kernel estimates of multivariated densities. Technometrics, 19: 135-144.

Bickel P. 1978. Using residuals robustly I: tests for heteroscedasticity, nonlinear. The Annals of Statistics, 6: 266-291.

Bickel P J, Klaassen A J, Ritov Y, Wellner J. A. 1993. Efficient and Adaptive Inference for Semiparametric Models. Baltimore: Johns Hopkins Univ Press.

Boor C A. 1978. A Practical Guide to Spline. New Work: Springer-Verlag.

Bowman A, Azzalini A. 1997. Applied Smoothing Techniques for Data Analysis. Oxford: Oxford University Press.

Breusch T S, Pagan A R. 1979. A simple test for heteroscedasticity and random coefficient variation. Econometrica, 47(5): 1287-1294.

Breusch T S, Pagan A R. 1980. The Lagrange multiplier test and its applications to model specification in econometrics. The Review of Econometric Studies, 47: 239-253.

Brunsdon C, Fotheringham A S, Charlton M. 1996. Geographically weighted regression: a method for exploring spatial nonstationarity. Geographical Analysis, 28: 281-298.

Buja A, Hastie T, Tibshirani R. 1989. Linear smoothers and additive models (with discussion). The Annals of Statistics, 17: 453-555.

Cai Z W, Hurvich C M, Tsai C L. 1998. Score tests for heteroscedasticity in wavelet regression. Biometrika, 85: 229-234.

Cai Z W, Fan J Q, Li R Z. 2000a. Efficient estimation and inference for varying coefficient model. Journal of the American Statistical Association, 95(451): 888-902.

Cai Z W, Fan J Q, Yao Q W. 2000b. Functional-coefficient regression models for nonlinear times series. Journal of the American Statistical Association, 95(451): 941-956.

Carroll R J, Ruppert D. 1981. On robust tests for heteroscedasticity. The Annals of Statistics, 9: 205-209.

Carroll R J, Ruppert D. 1988. Transformation and Weighting in Regression. London: Chapman and Hall.

Carroll R J, Fan J Q, Gijbels I, Wand M P. 1997. Generalized partially linear single index models. Journal of the American Statistical Association, 92: 477-489.

Carroll R J, Ruppert D, Stefanski L A, Crainiceanu C M. 2006. Measurement Error in Nonlinear Models. 2nd ed. New York: Chapman and Hall.

Cleveland W S. 1979. Robust locally weighted regression and smoothing scatterplots. Journal of the American Statistical Association, 74: 829-836.

Cleveland W S, Devlin S J. 1988. Locally weighted regreesion: an approach to regression analysis by local fitting. Journal of the American Statistical Association, 83: 596-610.

Chamberlain G. 1992. Efficiency bounds for semiparametric regression. Econometrica, 60: 567-596.

Chiou J M, Muller H G. 1999. Nonparametric quasi-likelihood. The Annals of Statistics, 27: 36-64.

Chen H. 1988. Convergence rates for parametric components in a partially linear model. The Annals of Statistics, 16: 136-146.

Chen R, Tsay S. 1993. Functional-coefficient autoregressive models. Journal of the American Statistical Association, 88: 298-308.

Cheng P E. 1994. Nonparametric estimation of mean functionals with data missing at random. Journal of the American Statistical Association, 89: 81-87.

Cheng C L, Van Ness J W. 1999. Statistical Regression with Measurement Error. London: Arnold.

Chow G C. 1960. Tests of Equality Between Sets of Coefficients in Two Linear Regressions. Econometrica, 28: 591-605.

Chiang C, Rice J A, Wu C. 2001. Smoothing spline estimation for varying coefficient models with repeatedly measured dependent variables. Journal of the American Statistical Association, 96: 605-619.

Chu C K, Cheng P E. 1995. Nonparametric regression estimation with missing data. Journal of Statist Planning Inference, 48: 85-99.

Cox D, Koh E, Wahba G, Yandell B S. 1988. Testing the (parametric) null model hypothesis in (semiparametric) partial and generalized spline models. The Annals of Statistics, 16: 113-119.

Crainiceanu C, Ruppert D, Claeskens G, Wand W P. 2005. Exact likelihood ratio tests for penalised splines. Biometrika, 92(1): 91-103.

Cui H J, Li R C. 1998. On parameter estimation for semi-linear errors-in-variables models. Journal of Multivariate Analysis, 64: 1-24.

Cui H J, Kong E F. 2006. Empirical likelihood confidence region for parameters in semi-linear errors-in-variables models. Scandinavian Journal of Statistics, 33: 153-168.

Dette H, Munk A. 1998. Testing heteroscedasticity in nonparametric regression. Journal of the Royal Statistical Sociation Series B, 60: 693-708.

Dette H. 1999. A consistent test for the functional form of a regression based on a difference of variance estimators. Annals of Statistics, 27: 1012-1040.

Dette H. 2000. On a nonparametric test for linear relationships. Statistics and Probability Letters, 42: 39-46.

Dette H, Neumeyer N. 2001. Nonparametric analysis of covariance . The Annals of Statistics, 29: 1361-1400.

Dierckx P. 1995. Curve and Surface Fitting with Splines (Monographs on Numerical Analysis). Oxford: Oxford University Press.

Donald G, Newey K. 1994. Series estimation of semilinear models. Journal of Multivariate Analysis, 50: 30-40.

Efromovich S. 1999. Nonparametric Curve Estimation. New York: Springer-Verlag.

Ellison G, Ellison S. 2000. A simple framework for nonparametric specification testing. Journal of Econometrics, 96: 1-23.

Engle R F, Granger W J, Rice J, Weiss A. 1986. Semiparametric estimates of the relation between weather and electricity sales. Journal of the American Statistical Association, 80: 310-319.

Eubank R L, Spiegelman C H. 1990. Testing the goodness of fit of a linear model via nonparametric regression techmques. Journal of the American Statistical Association, 85: 387-392.

Eubank R L, Thoms W. 1993. Detecting heteroscedasticity nonparametric regession. Journal of the Royal Statistical Sociation Series B, 55: 145-155.

Eubank R L. 1999. Nonparametric Regression and Spline Smoothing. New York: Marcel Dekker.

Eubank R L, Huang C, Maldondo Y. 2004. Smoothing spline estimation in varying coefficient models. Journal of the Royal Statistical Sociation Series B, 66(3): 653-667.

Fan J Q, Gijbels I. 1992. Variable bandwidth and local linear regession smoothers. The Annals of Statistics, 20: 2008-2036.

Fan J Q, Gibbers I. 1996. Local Polynomial Modeling and Its Applications. London: Chapman and Hall.

Fan J Q, Huang T. 2005. profile likelihood inferences on semiparametric varying-coefficient partially linear models. Bernoulli, 11: 1031-1057.

Fan J Q, Huang T, Li R Z. 2007. Analysis of longitudinal data with semiparametric estimation of covariance function. Journal of American Statistical Association, 102: 632-641.

Fan J Q, Jiang J C. 2005. Nonparametric Inferences for Additive Models. Journal of the American Statistical Association, 100: 890-907.

Fan J Q, Jiang J C. 2007. Nonparametric inference with generalized likelihood ratio tests (with discussion). Test, 16: 409-478.

Fan J Q, Li R Z. 2001. Variable selection via nonconcave penalized likelihood and its oracle properties. Journalof the American Statistical Association, 96: 1348-1360.

Fan J Q, Zhang C M, Zhang J. 2001. Generalized likelihood ratio statistics and Wilks phenomenon. The Annals of Statistics, 29: 153-193.

Fan J Q, Zhang W Y. 1999. Statistical estimation in varying-coefficient models. The Annals of Statistics, 27(5): 1491-1518.

Fan Y Q, Li Q. 2002. A consistent model specification test based on the kernel sum of squares of residuals. Econometric reviews, 21: 337-352.

Fan Y Q, Li Q. 2003. A kernel-based method for estimating additive partially linear models. Statistica Sinica, 13: 739-762.

Fan Y Q, Ullah A. 1999. Asymptotic Normality of a combined regression estimator. Journal of Multivariate Analysis, 71: 191-240.

Feng S Y, Xue L G. 2013. Bias-corrected statistical inference for partially linear varying coefficient errors-in-variables models with restricted condition. Annals of the Institute of Statistical Mathematics, in press.

Fox J. 2000. Nonparametric Simple Regression: Smoothing Scatterplots. Sage: Thousand Oaks.

Friedman J H, Stuetzle W. 1981. Projection pursuit regrssion. Journal of the American Statistical Association, 76: 817-823.

Fuller W A. 1987. Measurement Error Models. New York: Wiley.

Fung W K, Zhu Z Y, Wei B C, He X M. 2002. Influence diagnostics and outlier tests for semiparametric mixed models. Journal of the Royal Statistical Sociation Series B, 64: 565-579.

Gijbels I, Rousson V. 2001. Nonparametric least-squares test for checking a polynomial relationship. Statistics and Probability Letters, 51: 253-261.

Glejser H. 1969. A new test for heteroscedasticity. Journal of the American Statistical Association, 64: 316-323.

GonzGalez Manteiga W, Cao R. 1993. Testing hypothesis of general linear model using nonparametric regression estimation. Test, 2: 161-189.

Gonzalez-Manteiga, Aneiros-Perez G. 2003. Testing in partial linear regression models with dependent errors. Journal of nonparametric statistics, 15: 93-111.

Gozalo P, Linton O. 2000. Local nonlinear least squares: using parametric information in nonparametric regressin. Journal of Econometrics, 99: 63-206.

Green P J, Silverman B W. 1994. Nonparametric Regression and Generalized Linear Models. Vol.58 of Monographs on Statistics and Applied Probability, Chapman and Hall, London.

Gu C. 2002. Smoothing Spline ANOVA Models. New York: Springer-Verlag.

Gujarati D. 1970. Use of dummy variables in testing for equality between sets ofcoefficients in two linear regressions: a generalization. The American Statistician, 24: 50-52.

Hall P, Hart J D. 1990. Bootstrap test for difference between means in nonparametric regression. Journal of the American Statistical Association, 85: 1039-1049.

Hansen M H, Huang J Z. Kooperberg C, Stone C J, Truong Y K. 2003. Statistical Modelling with Spline Functions: Methodology and Theory. New York: Springer-Verlag.

Hardle W, Stoker T M. 1989. Investigating smooth multiple regression by the method of average derivatives. Journal of the American Statistical Association, 84: 986-995.

Hardle W. 1990. Applied Nonparametric Regression. Econometric Society Monographs No.19, Cambridge: Cambridge University Press.

Hardle W. 1991. Smoothing Techniques, With Implementations in S. New York: Springer.

Hardle W, Mammen E. 1993. Testing parametric versus nonparametric regression. The Annals of Statistics, 21: 1926-1947.

Hardle W, Kerkyacharian G, Picard D, Tsybakov A B. 1998. Wavelets, Approximation, and Statistical Applications. New York: Springer.

Hardle W, Liang H, Gao J. 2000. Partially Linear Models. Heidelberg: Springer, Physica-Verlag, Heidelberg.

Hardle W, Muller M, Sperlich S, Werwatz A. 2004. Nonparametric and Semiparametric Models. Berlin: Springer-Verlag.

Harrison M J, Mccabe B P M. 1979. A test for heteroscedasticity based on ordinary least squares residuals. Journal of the American Statistical Association, 74: 494-500.

Hart J D. 1997. Nonparametric Smoothing and Lack-of-Fit Tests. New York: Springer.

Hastie T J, Tibshirani R J. 1986. Generalized additive models (with discussion). Statistical Science, 1(2): 297-318.

Hastie T J, Tibshirani R J. 1990. Generalized Additive Models. Vol.43 of Monographs on Statistics and Applied Probability, London: Chapman and Hall.

Hastie T J, Tibshirani R J. 1993. Varying-coefficient models (with discussion) Journal of the Royal Statistical Sociation Series B, 55: 757-796.

Heckman N. 1986. Spline smoothing in a partially linear model. Journal of the Royal Statistical Sociation Series B, 48: 244-248.

Hjort N L, Jones M C. 1996. Locally parametric nonparametric density estimation. The Annals of Statistics, 23: 882-904.

Hoover D R, Rice J A, Wu C O, Yang L P. 1998. Nonparametric smoothing estimation of time-varying coefficient models with longitudinal data. Biometrika, 85(4): 809-822

Horowitz, J. L. 1998. Semiparametric Methods in Econometrics. New Work: Springer.

Hu Z H, Wang N, Carroll R J. 2004. profile-Kernel Versus backfitting in the partially linear models for longitudinal/clustered data. Biometrika, 91: 251-262.

Hu X M, Wang Z Z, Zhao Z Z. 2009. Empirical likelihood for semiparametric varying-coefficient partially linear errors-in-variables models. Statistics and Probability Letters, 79: 1044-1052

Huang J Z, Wu C O, Zhou L. 2002. Varying-coefficient models and basis function approximations for the analysis of repeated measurements. Biometrika, 89: 111-128.

Huang J Z, Wu C O, Zhou L. 2004. Polynominal spline estimation and inference for varying coefficient models with longitudinal data. Statistic Sinica, 14: 763-788.

Imhof J P. 1961. Computing the distribution of quadratic forms in normal variables. Biometrika, 48: 419-426.

Ip W, Wong H, Zhang R Q. 2007. Generalized likelihood ratio tests for varying-coefficient models with different smoothing variables. Computational Statistics Data Analysis, 51: 4543-4561.

Jayasuriya B R. 1996. Testing for polynomial regression using nonparametric regression techniques. Journal of the American Statistical Association, 91: 1626-1631.

Jiang J C, Zhou H, Jiang, X J, Peng J. 2007. Generalized likelihood ratio tests for the structures of semiparametric additive models. The Canadian Journal of Statistics, 35: 381-398.

Johnson N L, Kotz A B. 1970. Continuous Univariate Distribution. New York: John Wiley and Sons.

Keilegom I V, Wang L. 2010. Semiparametric modeling and estimation of heteroscedasticity in regression analysis of cross-sectional data. Electronic Journal of Statistics, 4: 133-160.

King E C, Hart J D, Wehrly T E. 1991. Testing the equality regression curves using linear smoothers. Statistics and Probability Letters, 12: 239-247.

Kolaczyk E D. 1994. Empirical likelihood for generalized linear models. Statistica Sinica, 4: 199-218.

Li Q, Ullah A. 1998. Estimating partially linear panel data models with one-way error components. Econometric Reviews, 17(2): 145-166.

Li, Q, Wang S. 1998. A simple consistent bootstrap test for a parametric regression function. Journal of Econometrics, 87: 145-165.

Li Q. 2000. Efficient estimation of additive partially linear models. International Economic Review, 41: 1073-1092.

Li L, Greene T. 2008. Varying coefficients model with measurement error. Biometrics, 64: 519-526.

Liang H, Hardle W, Carroll R J. 1999. Estimation in a semiparametric partially linear errors-in-variables model. The Annals of Statistics, 27: 1519-1535.

Liang H, Wang S, Robins J M, Carroll R J. 2004. Estimation in Partially linear models with missing covariates. Journal of the American Statistical Association, 99(466): 357-367.

Liang H. 2006. Estimation in Partially linear models and numerical comparisons. Computational Statistics Data Analysis, 50: 675-687.

Linton O, Nielsen J P. 1995. A kernel method of estimating structured nonparametric regression based on marginal integration. Biometrika, 82: 93-101.

Liang H, Wang S J, Carroll R J. 2007. Partially linear models with missing response variables and error-prone covariates. Biometrika, 94: 185-198.

Liang H, Thurston H, Ruppert D, Apanasovich T. 2008. Additive partial linear models with measurement errors. Biometrika, 95: 667-678.

Liang H, Thurston H, Ruppert D, Apanasovich T. 2008. Additive partial linear models with measurement errors. Biometrika, 95: 667-678.

Liang H, Su H Y, Thurston S. 2009. Empirical likelihood based inference for additive partial linear measurement error models. Statistics and Its Interface, 2: 83-90.

Linton O, Hardle W. 1996. Estimation of additive regression models with known links. Biometrika, 83(3): 529-540.

Linton O. 1997. Efficient estimation of additive nonparametric regression models. Biometrika, 82: 529-540.

Little R J A, Rubin D B. 2002. Statistical Analysis With Missing Data. 2nd ed. New York: John Wiley.

Liu Z, Stengos T, Li Q. 2000. Nonparametric model check based on local polynomial fitting. Statistics and Probability Letters, 48: 327-334.

Loader C. 1999. Local Regression and Likelihood. New York: Springer.

Louis A K, Maass D, Rieder A. 1997. Wavelets: Theory and Applications. New York: Wiley.

Ma Y Y, Chiou J M, Wang N. 2006. Effcient semiparametric estimator for heterocesdastic partially linear models. Biometrika, 943: 75-84.

Mack Y P, Silverman B W. 1982. Weak and strong uniform consistency of kernel regression estimates. Z. Wahrsch. Verw. Gebiete, 61: 405-415.

Manzana S, Zeromb D. 2005. Kernel estimation of a partially linear additive model. Statistics and Probability Letters, 72: 313-322.

Miles D, Mora J. 2003. On the performance of nonparametric specification tests in regression models. Computational Statistics and Data Analysis, 42: 477-490.

Muller H G, Stadtmuller U. 1987. Estimation of Heteroscedasticity in Regression Analysis, The Annals of Statistics, 15: 610-625.

Muller H G, Zhao P L. 1995. On a semiparametric variance function model and a test for heteroscedasticity. The Annals of Statistics, 23: 946-967.

Muller H G. 1988. Nonparametric Regression Analysis of Longitudinal Data. Lecture Notes in Statistics, New York: Springer-Verlag.

Nadaraya E A. 1989. Nonparametric Estimation of Probability Densities and Regression Curves. Kluwer: Boston.

Neyman J. 1959. Optimal asymptotic test of composite statistical hypothesis//Grenander U. Probability and Statistics, the Harald Cramer Volume. Almqvist and Wiksell, Uppsala, 213-234.

Ogden R T. 1996. Essential Wavelets for Statistical Applications and Data Analysis. Birkhauser, Boston.

Opsomer J D, Ruppert D. 1997. Fitting a bivariate additive model by local polynomial regression. The Annals of Statistics, 25: 186-211.

Opsomer J D, Ruppert D. 1999. A root-n consistent backfitting estimator for semiparametric additive modelling. Journal of Computational Graphical of Statistics, 8: 715-732.

Opsomer J D. 2000. Asymptotic properties of backfitting estimators. Journal of Multivariate Analysis, 73: 166-179.

Owen A B. 1988. Empirical likelihood ratio confidence intervals for single functional. Biometrika, 75: 237-249.

Owen A B. 1990. Empirical likelihood ratio confidence regions. The Annals of Statistics, 18: 90-120.

Owen A B. 1991. Empirical likelihood for linear models. The Annals of Statistics, 19: 1725-1747.

Pagan A, Ullah A. 1999. Nonparametric Econometrics (Themes in Modern Econometrics). Cambridge: Cambridge University Press.

Pearson E S. 1959. Note on an approximation to the distribution of non-central χ^2. Biometrika, 46: 364.

Przystalski M, Krajewski P. 2007. Constrained estimators of treatment parameters in semiparametric models. Statistics and Probability Letters, 77(10): 914-919.

Qin J, Lawless J. 1994. Empirical likelihood and general estimating equations. The Annals of Statistics, 22: 300-325.

Rao C R. 1973. Linear Statistical Inference and Its Application. New York: Wiley.

Rao C R, Toutenburg H. 1999. Linear Models: Least Squares and Alternatives. 2nd. Berlin: Springer.

Rao C R, Toutenburg H, Shalabh, Heumann C. 2008. Linear Models: Least Squares and Alternatives. Berlin: Springer.

Robinson P, 1988. Root-N-consistent semiparametric regression. Econometrica, 56: 931-954.

Ruppert D, Wand M P, Carroll R J. 2003. Semiparametric Regression. Cambridge: Cambridge University Press.

Ruppert D, Wand M P, Host U, Hossjer O. 1997. Local polynomial variance-function estimation. Technometrics, 39: 262-273.

Ryen T P. 1997. Modern Regression Methods. New York: Wiley.

Schimek M G. 2000. Smoothing and Regression. New York: Wiley.

Schumaker L L. 1981. Spline Function. New York: Wiley.

Scott D W. 1992. Multivariate Density Estimation: Theory, Practive and Visualization. New York: Wiley.

Seber G A F, Wild C J. 1989. Nonlinear Regression. New York: Wiley.

Severini T, Staniswalis J. 1994. Quasi-likelihood estimation in semiparametric models. Journal of the American Statistical Association, 89: 501-511.

Shalabh, Garg G, Misra N. 2007. Restricted regression estimation in measurement error models. Computational Statistics and Data Analysis, 52: 1149-1166.

Shi J, Lau T S. 2000. Empirical likelihood for partially linear models. Journal of Multivariate Analysis, 72: 132-149.

Silverman B W. 1986. Density Estimation for Statistics and Data Analysis. Vol.26 of Monographs on Statistics and Applied Probability, Chapman and Hall, London.

Simonoff J. 1996. Smoothing Methods in Statistics. New York: Springer.

Solomon H, Stephens M A. 1977. Distribution of a sum of weighted chisquare variables. Journal of the American Statistical Association, 72: 881-885.

Speckman P. 1988. Kernel smoothing in partial linear models. Journal of the Royal Statistical Sociation Series B, 50: 413-436.

Stuart A, Ord D K. 1994. Kendall 's Advanced Theory of Statistics. Vol 1, Distribution Theory (6th edition). London: Edward Arnold.

Tarter M F, Lock M D. 1993. Model-free Curve Estimation. New York: Chapman and Hall.

Thompson J R, Tapia R A. 1990. Nonparametric Function Estimation: Modeling and Simulation. Philadelphia: SIAM.

Toutenburg H. 1982. Prior Information in Linear Models. New York: Wiley.

Ullah A. 1985. Specification analysis of econometric models. Journal of Quantitative Economics, 1: 187-210.

Ullah A, Vinod H D. 1988. General Nonparametric regression estimation and testing in econometircs//Maddala G S, Rao C R, Vinod H D. eds. Handbook of Statistics, Amsterdam, Elsevier Publishing Co, (11): 85-116.

Wahba G. 1990. Spline models for observational data. SIAM, Philadelphia, Pennsylvania.

Walter G G, Shen X. 2001. Wavelets and Other Orthogonal Systems. 2nd ed. Boca Raton: Chapman and Hall/CRC Press.

Wand M P, Jones M C. 1995. Kernel Smoothing. London: Chapman and Hall.

Wang Q H, Linton O, Hardle W. 2004. Semiparametric regression analysis with missing response at random. Journal of the American Statistical Association, 99(466): 334-345.

Wang Q H. 1999. Estimation of partial linear error-in-variables models with validation data. Journal of Multivariate Analysis, 69: 30-64.

Wang Q H, Rao J N K. 2002. Empirical Likelihood-based inference in linear errors-in-covariables models with validation data. Biometrika, 89: 345-358.

Wang Q H, Sun Z H, 2007. Estimation in partially linear models with missing responses at random. Journal of Multivariate Analysis, 98: 1470-1493.

Wang X L, Li G R, Lin L. 2011. Empirical likelihood inference for semi-parametric varying-coefficient partially linear EV models. Metrika, 73: 171-185.

Wei C H, Jia X J, Hu H S. 2013a. Restricted estimation in partially linear errors-in-variables models. Communication in Statistics: Simulation and Computation, 42(8): 1836-1847.

Wei C H, Jia X J, Hu H S. 2013b. Statistical inference on partially linear additive models with missing response variables and error-prone covariates. Communication in Statistics: Theory and Methods, accepted.

Wei C H, Liu C L. 2012. Statistical inference on semiparametric partially linear additive models. Journal of Nonparametric Statistics, 24: 809-823.

Wei C H, Luo Y B, Wu X Z. 2012. Empirical likelihood for partial linear additive errors-in-variables models. Statistics Papers, 53: 485-496.

Wei C H, Mei C L. 2012. Empirical likelihood for partially linear varying-coefficient models with missing response variables and error-prone covariates. Journal of the Korean Statistical Society, 41: 97-103.

Wei C H, Wan L J, Liu C L. 2013. Efficient estimation in heteroscedastic partially linear varying coefficient models. Communication in Statistics: Simulation and Computation, accepted.

Wei C H, Wang Q H. 2012. Statistical inference on restricted partially linear additive errors-in-variables models. Test, 21: 757-774.

Wei C H, Wu X Z. 2008. Asymptotic normality of estimators in partially linear varying coefficient models. Journal of Mathematical Research and Exposition, 28: 877-885.

Wei C H. 2012a. Statistical inference on restricted partially linear varying coefficient errors-in-variables models. Journal of Statistical Planning and Inference, 142: 2464-2472.

Wei C H. 2012b. Statistical inference in partially linear varying-coefficient models with missing responses at random. Communication in Statistics: Theory and Methods, 41(7): 1284-1298.

Wu C O, Chiang C T, Hoover D R. 1998. Asymptotic confidence regions for kernel smoothing of a varying-coefficient model with longitudinal data. Journal of the American Statistical Association, 93: 1388-1402.

Wu C O, Chiang C T. 1998. Kernel smoothing on varying coefficient models with longitudinal dependent variable. Statistic Sinica, 10: 433-456.

Xia Y C, Li W K. 1999. On the estimating and testing of functional-coefficient linear models. Statistica Sinica, 9: 737-757.

Xia Y C, Zhang W Y, Tong H. 2004. Efficient estimation for semivarying-coefficient models. Biometrika, 91: 661-681.

Yanagimoto T, Yanagimoto M. 1987. The use of marginal likelihood for a diagnostic test for the goodness of fit of the simple linear regression model. Technometrics, 29: 95-101.

Yatchew A. 1997. An elementary estimator of the partial linear mdoel. Economics Letters, 57: 135-143.

Yatchew A. 1999. An elementary nonparametric differencing test of equality of regression functions. Economics Letters, 62: 271-278.

Yatchew A. 2003. Semiparametric Regression for the Applied Econometrician. Cambridge: Cambridge University Press.

You J H, Chen G M. 2005. Testing heteroscedasticity in partially linear regression models. Statistics and Probability Letters, 73: 61-70.

You J H, Chen G M. 2006. Estimation of a semiparametric varying-coefficient partially linear errors-in-variables model. Journal of Multivariate Analysis, 97: 324-341.

You J H, Zhou Y. 2006. Empirical likelihood for semiparametric varying-coefficient partially linear regression models. Statistics and Probability Letters, 76: 412-422.

You J H, Zhou Y, Chen G M. 2006. Corrected local polynomial estimation in varying coefficient models with measurent errors. Canadian Journal of Statistics, 34: 391-410.

Young S G, Bowman A W. 1995. Non-parametric analysis of covariance. Biometrics, 51: 920-931.

Zhang C M, Dette H, 2004. A power comparison between nonparametric regression tests. Statistical Probability Letters, 66: 289-301.

Zhang D W, Lin X H, Raz J, Sowers M. 1998. Semiparametric stochastic mixed models for longitudinal data . Journal of the American Statistical Association, 93: 710-719.

Zhang R Q, Huang Z S. 2009. Statistical inference on parametric part for partially linear single-index model. J. Science in China(A), 52: 2227-2242.

Zhang W W, Li G R, Xue L G. 2011. profile inference on partially linear varying-coefficient errors-invariables models under restricted condition. Computational Statistics and Data Analysis, 55: 3027-3040.

Zhang W Y, Lee S, Song X. 2002. Local polynomial fitting in semivarying-coefficient model. Journal of Multivariate Analysis, 82(1): 166-188.

Zhang X Y, Liang H, 2011. Focused information criterion and model averaging for generalized additive partial linear models. The Annals of Statistics, 39: 174-200.

Zheng J X. 1996. A consistent test of a functional form via nonparametric estimation techniques. Journal of Econometrics, 75: 263-289.

Zhou X, You J H. 2004. Wavelet estimation in varying coefficient partially linear regression models. Statistics and Probability Letters, 68: 91-104.

索　　引